D0566992

Natural Gas Imports from the Soviet Union

Joseph T. Kosnik

The Praeger Special Studies program—utilizing the most modern and efficient book production techniques and a selective worldwide distribution network—makes available to the academic, government, and business communities significant, timely research in U.S. and international economic, social, and political development.

Natural Gas Imports from the Soviet Union

Financing the North Star Joint Venture Project

Univarsitas
BIBLIOTHECA
Ottawiensis

Praeger Publishers   New York   Washington   London

PRAEGER SPECIAL STUDIES IN INTERNATIONAL ECONOMICS AND DEVELOPMENT

008902

Library of Congress Cataloging in Publication Data

Kosnik, Joseph T
 Natural gas imports from the Soviet Union.

 (Praeger special studies in international
economics and development)
 Bibliography: p.
 Includes index.
 1. Gas, Natural—Russia. 2. Joint adventures—Russia.
3. Gas, Natural—United States. I. Title.
HD9581. R92K68   382'. 42'2850973   75-23976
ISBN 0-275-01480-0

HD
9581
. R92 K68
1975

PRAEGER PUBLISHERS
111 Fourth Avenue, New York, N.Y. 10003, U.S.A.

Published in the United States of America in 1975
by Praeger Publishers, Inc.

All rights reserved

© 1975 by Praeger Publishers, Inc.

Printed in the United States of America

This book is dedicated to my wife Joan and children
Tom, Jean, Mark, Julie and Chris.

FOREWORD
Philip D. Grub

The United States is at an historic cross-road in economic and trade relations, and is confronted with decisions which will have a major impact on the U.S. and world economies for years to come. One decision involves trade relations with the Soviet Union and, indirectly, trade with all Communist nations, particularly those in the eastern bloc.

In the current atmosphere of detente, trade assumes a position of major importance because it is the one area of mutual interest which the Communist nations are anxious to exploit and non-communist nations need for economic growth; consequently, it provides the wherewithal for peaceful interchange between potentially hostile antagonists. The problems of trade do not reduce to simple economic or rational considerations involved, but provide a realm of political and ideological imponderables.

The author of this book, Joseph T. Kosnik, has done a thorough job of integrating the economic, financial and rational factors with the institutional, ideological and psychological considerations inherent in U.S.-Soviet trade. He explodes myths and misconceptions by pointing out the real apprehensions which underlie the joint venture proposals for financing such trade. In the specific areas of Soviet natural gas resources, energy, gold reserves and hard currency trade earnings, the data presented are the best available in current literature. The book is a serious work in that it presents technical data and accurate statistics while written in readable prose without technical jargon. Key trade considerations are made strongly and in straightforward, understandable language. The principles are broad and extend far beyond financing trade or the North Star joint venture case study. The book is essential reading for all engaged in U.S.-Soviet relations—academician, businessman, and government decision-maker alike.

PREFACE

The Soviet Union is generously endowed with raw materials and energy resources, but it lacks investment funds, technology, and managerial know-how in order to extract them efficiently. Also, agriculture has been a perennial Soviet problem. The United States enjoys a bounty in agricultural products and leads the world in technology and management; however, its economy suffers from a shortage of energy resources. Rationally and economically, the two nations are trade compatible. Unfortunately, political, ideological, and institutional constraints have created obstacles to trade between them.

It would be extremely difficult to evaluate all Soviet proposals for barter trade* or industrial and commercial cooperation and to devise financial arrangements applicable in all cases. Therefore, the scope of this study is limited to an analysis of the Soviet "North Star" project. The project concerns the movement of liquefied natural gas (LNG) from western Siberia to the U.S. East Coast port of Philadelphia and involves gas wells, pipelines, liquefaction plants, and LNG tankers. Included in the scope of this study is an evaluation of the proposed joint venture arrangement as a means of financing the North Star project and an outline of the problems encountered in dealing with governmental and business barriers in the United States and the Soviet Union as they concern the legal, technical, and institutional aspects of the joint proposal.

Specifically, this study examines the feasibility of exchanging USSR natural gas for U.S. technical and managerial know-how under the special financial arrangements of a joint venture. The United States would participate in the development of Soviet natural gas resources in exchange for payment in the form of production from the developed resources. Ideological-institutional constraints of the planned and free-market systems are analyzed. USSR natural gas reserves as well as production and consumption trends are projected, and Soviet technical and managerial needs are determined in order to evaluate the availability of natural gas for export. The investment risk or the cost of U.S. services is compared against the Soviet ability to pay in the form of gas exports or in gold or hard-currency trade surplus earnings.

----

*Barter trade is used in this study to mean an exchange of commodities and services through a price mechanism and is not limited to a strict barter exchange.

Several limitations are prevalent in the study. First, U.S.-Soviet trade is currently receiving much attention from government and business interests in America. Much of the information concerning specific agreements is proprietary and is held closely for security and business reasons. Approaches to government and banking officials in the United States revealed a willingness to discuss commercial relations in a general way, but a reluctance to provide primary data or to be quoted. Soviet documentary sources often obscure details, and data are frequently published for political or psychological effect. Therefore, the research required careful analysis and interpretation of the data that were available.

Second, the U.S. energy shortage arising from increasingly insufficient production from domestic sources for the period through the 1980s is accepted as a fact. Although the view has been expressed that the shortage was caused by business manipulation and government bungling, there seems to be no disagreement that America is confronted by a serious domestic energy shortage. How to solve the energy problem is a question far beyond the limits of this book. Several studies have been conducted by the Tenneco group, however, which determined that natural gas delivered to Philadelphia from the Soviet Union at a cost of $1.25 per million British thermal units (BTUs) would be competitive with other energy alternatives.[1] Since the subject of this book was the analysis of investments in the Soviet Union, the Tenneco finding that natural gas imported from the Soviet Union would be cost competitive in the U.S. market was accepted, and comparisons with alternative sources of gas are shown. Imports of LNG require approval of the Federal Power Commission (FPC), and for this approval the price must be cost competitive.

Third, dependence on imports from the Soviet Union for energy resources involves the problem of national security considerations. A proper solution to this problem requires detailed analysis of available energy resources and consumption trends, but this type of analysis is not within the scope of this study. On the basis of trend projections, it is estimated that North Star would satisfy 0.6 percent of the total energy requirements of the United States, 2.5 percent of its total natural gas consumption, or 10 percent of the natural gas requirements of the northeastern geographic sector in the mid-1980 period.[2] Although loss of this resource would have repercussions, statements by official policy makers of the United States that the security of the nation would not be seriously damaged have been accepted.[3]

Although this book was conducted as a case study of the North Star project, the findings are believed to be widely applicable to other Soviet proposals. The Soviet invitation to participate in construction of the Kama River project and in the ammonia urea plants as well as the future invitation to bid on aluminum plants, cellulose and paper mills, and Amur-Baikal Railroad follow patterns which are familiar to the ones presented in the North Star project.

A study of this magnitude would be impossible without the assistance of many people. I accept full responsibility for all conclusions reached, but wish to acknowledge and express my gratitude to those who have helped. To the George Washington University, and particularly to Professors Phillip D. Grub and John P. Hardt, I am indebted for inspiration and encouragement to continue my study; to the Industrial College of the Armed Forces I am indebted for the excellent graphics and reproduction assistance; and to Mrs. Dorothy Couture, my patient and outstandingly professional typist, I am indebted for complete administrative and clerical assistance. To my family I am grateful for their patience, and to the many helpers too numerous to mention, I extend everlasting gratitude.

NOTES

1. North Star Project Feasibility Study, by Brown and Root, Inc., Tenneco Inc., and Texas Eastern Transmission Corporation, vol. 1: Text (Houston, March 31, 1972), p. 6.

2. Jack H. Ray, "The North Star Soviet Liquefied Natural Gas (LNG) Import Project," statement in U.S. Congress, House, Committee on Banking and Currency, International Economic Policy, Hearings before Subcommittee on International Trade on H.R. 774, H.R. 13838, H.R. 13839, H.R. 13840, 93rd Cong., 2nd sess., 1974, p. 229. (Hereafter cited as House Subcommittee Hearings on International Economic Policy, 1974.)

3. John N. Nassikas, chairman, Federal Power Commission, "The Role of Liquefied Natural Gas in U.S. Energy Policy," remarks before the Fourth International Conference on Liquefied Natural Gas at Palace of Nations, Algeria, June 24, 1974; and William J. Casey, statement before the Subcommittee on International Trade, in House Subcommittee Hearings on International Economic Policy, 1974, p. 697.

CONTENTS

LIST OF TABLES

LIST OF FIGURES AND MAPS

LIST OF ABBREVIATIONS

AGA	American Gas Association
AGIP	Assienda Generale Italiana Petroli
BCF	billion cubic feet
BCFD	billion cubic feet daily (or per day)
BCM	billion cubic meters
BTU	British thermal units
FPC	Federal Power Commission
GATT	General Agreement on Tariffs and Trade
GNP	gross national product
GOELRO	the Plan of the State Commission for Electrification of Russia
GOSPLAN	State Planning Commission
IBM	International Business Machines
LNG	liquefied natural gas
MCF	million cubic feet
MFN	most favored nation
NEP	New Economic Policy
NGRS	National Gas Reserves Study
NIOC	National Iranian Oil Company
ONGC	Oil and Natural Gas Commission
SNG	synthetic natural gas
TCF	trillion cubic feet
TCM	trillion cubic meters

INTRODUCTION
AND BACKGROUND

Before embarking on the complex study of financing the development of a Soviet gas field in Siberia and delivering its output to Philadelphia, it was necessary to conduct research to determine what investment practices had been used in other areas of the world and to identify as precisely as possible all elements of the North Star project. In addition, guidelines had to be established for the orderly conduct of the study.

One of the most recent techniques employed by the Soviet Union to attract the investments of foreign countries is the use of the joint venture arrangement. The origins of current Soviet joint venture proposals to the United States can be traced back to May 1958, when Soviet Premier Nikita Khrushchev called upon the West for a more rapid development of the Soviet chemical industry. In answer to his call, Western Europe, particularly West Germany, responded favorably and increased trade with the Union of Soviet Socialist Republics, but political, ideological, and institutional obstacles prevented a U.S.-Soviet accommodation. However, the natural gas shortage in the United States, which began to experience supply shortfalls in 1968, caused American businessmen to examine seriously the possibility of tapping the extensive Soviet natural gas reserves.

When Richard M. Nixon assumed the office of president of the United States in 1969, his policy of political-economic linkages gradually improved U.S.-Soviet relations. By 1971, proposals for major investments by the United States were categorized in four broad areas:

1. development of gas deposits in Siberia
2. development of the Udokan copper reserves
3. construction of the Kama River truck plant
4. modernization of Soviet agriculture and consumer industries.

In May 1972, President Nixon and Secretary-General Leonid Brezhnev agreed on the basic principles for improved Soviet-American trade, and subsequently, a three-year trade agreement was signed on October 18, 1972. The agreement proposed, as a minimum, that the bilateral trade of the previous three years be tripled, to reach a total turnover in the forthcoming three-year period of at least $1.5 billion.[1]

The trade agreement was partially implemented, even before enabling legislation was passed by the Congress, and resulted in a considerable increase in U.S.-Soviet trade. Total trade turnover increased from $219 million in 1971 to $1.4 billion in 1973,[2] exceeding the quantitative goals of the agreement. The enabling legislation passed by Congress in January 1975 contained restricting clauses, which caused the Soviets to deny formal implementation of the trade agreement.

Included in expanded U.S.-Soviet commercial relations is the joint venture arrangement, designed to promote expansion of U.S.-Soviet trade. Under the joint venture arrangement, advance production payments in the form of Western equipment, technology, and services are used for the purpose of constructing Soviet factories or development of Soviet natural resources. These Western investments are exchanged for a share of the future output of the Soviet enterprises created or natural resources developed. This exchange amounts to barter trade, except that a pricing mechanism is required to implement trade terms over time. The arrangement has become so common in the USSR that provisions for it are made in the annual economic plan for the development of the Soviet economy.[3]

In response to recent Soviet joint venture proposals, agreements of intent involving two huge projects in the Soviet Union have been signed. In one of the projects, a combination of Japanese private companies and government agencies together with three U.S. private corporations—Occidental Petroleum Corporation, El Paso Natural Gas Company, and Bechtel Incorporated—are analyzing the feasibility of moving between 2.0 and 2.4 billion cubic feet (BCF) of natural gas daily from Yakutsk, Siberia, to equally divided markets in Japan and on the U.S. West Coast. Only very preliminary studies on the project have been conducted so far, with more definitive engineering, economic, and financing studies planned for the near future. Funds to pay for the foreign goods and services needed to develop Soviet natural resources are expected to be made available through export-import credits and commercial financial sources. Credits are to be arranged through the combined efforts of the United States, Japan, and possibly a third country.

The other joint venture project, known as the North Star, is the subject of this study. It is being undertaken by a consortium comprised of Tenneco Inc., Texas Eastern Transmission Corporation, and Brown and Root and involves the movement of 2.1 BCF of natural gas daily, covering a period of 25 years, from the Urengoi field in Western Siberia to the Philadelphia, Pennsylvania, area. Details, which would permit the signing of firm contracts, are under negotiation. Long-term financing has surfaced as a major problem in the negotiations. Experts in international financing indicate that the North Star project is economically feasible, even though loans to the Soviet Union as large as $3 to $4 billion would be required. Of this amount, U.S.-Soviet negotiators expected to obtain at least a $1-billion loan from the U.S. Export-Import Bank together with a $1-billion loan guarantee.4 Restrictions included in the legislation passed by Congress appear to preclude this possibility. A final agreement between American and Soviet negotiators has not been signed.

PURPOSE OF THE STUDY

The purpose of this study is to explore the problems inherent in a joint venture project with the Soviet Union. The term "joint venture" is used in the same sense as that employed by the Soviets and is not meant to imply equity participation, as is common in Western usage. Further, the study evaluates the feasibility of using a joint venture arrangement to finance U.S. investment in the North Star project. Investment is used in a broad sense of providing a flow of income over a long period of time. Soviet law precludes private ownership of the means of production. Therefore, equity participation in the accepted Western practice is illegal in the Soviet Union. The Soviets use the term "investment," however, to mean advance production payments in connection with joint ventures.

One characteristic of U.S.-Soviet commercial relations during the current period of detente is the proliferation of confusing information regarding both countries. An attempt is made in this study, therefore, to gather data and analyze the conflicting facts which abound on the subject. This task, by itself, is designed as a service to the academic and business communities, in the hope that it will point out problem areas which arise in the forming of joint ventures.

RESEARCH QUESTION AND SUBSIDIARY QUESTIONS

The primary research question in this study is: Is the proposed joint venture arrangement a feasible model for U.S. participation in the North Star project? The question will be answered by analyzing the various aspects of the proposed exchange of Soviet raw materials for U.S. goods and services as a means of effecting significant increases in trade and by determining the feasibility of financing such a huge undertaking through a joint venture arrangement. In this sense, the approach is primarily deductive.

In addition to the primary question, several subsidiary questions are considered. They are

1. What arrangements are used in other parts of the world to finance exploration and trade in natural gas?
2. Does the U.S. natural gas shortage justify participation in North Star, and would the Soviet gas be cost competitive in the U.S. market?
3. Are Soviet natural gas reserves adequate to support North Star?
4. Do Soviet gas production and consumption trends permit the allocation of sufficient gas to North Star to assure deliveries on a reliable basis?
5. Is North Star feasible from the point of view of the Soviet balance of payments and debt service?
6. Will U.S. law permit export of the technology required?
7. Do U.S. tariff dumping regulations affect North Star?
8. What are the obstacles to the financing of North Star in the U.S. capital market?
9. What alternative sources are available to provide financing for North Star?

RESEARCH METHODOLOGY

In order to conduct this study, standard research methods were utilized. The initial step was a review of available literature concerning joint ventures. It was discovered that because the concept of current joint ventures, as used by the Soviet Union, is comparatively recent, relatively little literature on the subject exists. From the literature that was available, however, it was possible to reconstruct a brief history and background of joint ventures, which is found in Chapter 2.

The next step was the gathering of documentary material pertaining to the North Star project and its analysis. The collecting of necessary material was not always easy, as the parties involved in the project were reluctant to disclose information.

Interviews were then scheduled with American officials involved in negotiations or responsible for government policy and business investment decisions. Interviews with Soviet officials were limited to meetings and discussions with members of the Soviet Embassy in Washington, D.C. Soviet officials were helpful and cooperative and provided referrals to official publications or press releases of the

Soviet government. The research lacks the detail which might have been provided by interviews with Soviet officials in Moscow, but the factual data extracted from official Soviet documents are believed to be accurate.

One of the problems encountered in evaluating the information available was that of translating concepts from Russian to English and vice versa and ensuring that the concepts are accurately portrayed. An additional problem was that often, in both languages, the same words were used to convey somewhat different meanings.

Documents utilized in the study included official U.S. and Soviet agreements, regulations, decrees, studies, publications, and press releases, which were analyzed and supplemented by a review of books, journals, and press articles. Banking and corporate data were analyzed, and interviews were scheduled with banking, corporate, and governmental officials in order to clarify controversial points.

Interviews conducted in the United States included private and government bank officials, corporate executives, government officials, and consultants. Many of those interviewed asked that their opinions and information be kept confidential. The analyses and findings of this study were greatly influenced and assisted by the interviews, though credit could not be given to the sources. The findings are the sole responsibility of the researcher.

The Russian language was used as a research tool. Soviet data concerning petroleum resources are limited because, under Soviet law, they are considered state secrets. Natural gas, however, is not classified as a secret resource, and extensive coverage of the subject occurs in the Soviet press and periodicals. Soviet data required careful analysis because the measurement and definition of resources differ from that used in the United States. Also, political and psychological factors are frequently injected in the publication of data in the Soviet Union. Quantitative data on natural gas are presented in cubic meters for the Soviet Union and in cubic feet for the United States in order to preserve source accuracy. For conversion purposes, 1 cubic meter is equal to 35.314 cubic feet of natural gas.

ORGANIZATION OF THE STUDY

The study is divided into three parts: I—introduction and background, II—institutional obstacles within the Soviet system, and III—institutional obstacles within the United States.

Chapter 1 contains general background information. In Chapter 2, the joint venture arrangement is defined, and a brief history and background of joint ventures are given. Current joint ventures, as used by the Soviet Union, are a recent innovation; hence little literature on the subject is available.

In Chapter 3, an analysis is made of the current shortage of natural gas in the United States and the need for imports of foreign gas. In Chapter 4, the efforts of Tenneco Inc. to develop sources of gas supplies are presented, and a description of the chief components of the North Star project are included.

In Chapter 5, the Soviet governmental structure, legal base, and ideological obstacles are analyzed. In Chapter 6, the feasibility of North Star is determined by analyzing the adequacy of gas reserves and the production and consumption trends for natural gas in the Soviet economy. In Chapter 7, the state of Soviet technology is assessed and needs of gas technology analyzed. In Chapter 8, the credit worthiness of the Soviet Union is evaluated in terms of Soviet hard-currency earnings, gold reserves, and debt service ratio.

In Chapter 9, the trade authorizing factors, that is, most-favored-nation (MFN) treatment, dumping, and technology, are analyzed to identify obstacles to North Star within the U.S. system. In Chapter 10, the trade financing factors, that is, Export-Import Bank financing, commercial bank financing, and limitations related to debt default, are analyzed to ascertain obstacles to financing North Star in the U.S. capital market.

In Chapter 11, answers to the research question and subsidiary questions are given, and conclusions regarding the feasibility of using a joint venture arrangement to finance North Star are presented.

NOTES

1. U.S. Department of Commerce, Domestic and International Business Administration, Bureau of East-West Trade, U.S.-Soviet Commercial Agreements, 1972: Texts, Summaries and Supporting Papers (Washington, D.C.: Government Printing Office, 1973), p. 89.

2. U.S. Department of Commerce, Domestic and International Business Administration, Bureau of East-West Trade, East-West Trade, export administration report for fourth quarter, 1973 (Washington, D.C.: Government Printing Office, 1974), pp. 59-64.

3. N. K. Baibakov, "O Gosudarstvennom Plane Razvitia Narodnovo Khozyaistva SSSR na 1974 God; Vneshniye Ekonomicheskiye Svyazi" [Concerning the Plan for Development of the USSR National Economy in 1974, External Economic Ties], Pravda, December 13, 1973, p. 3.

4. Information in this paragraph was compiled from a series of interviews and correspondence with Jack H. Ray, president of the Tennessee Gas Transmission Company, who is the originator, planner, and coordinator of the project for Tenneco Inc., and Dan Walsh, manager of LNG projects, Texas Eastern Transmission Corporation, during the period September 1973 through July 1974.

2

**BRIEF BACKGROUND
AND HISTORY OF
JOINT VENTURES**

In order to understand the significance and historical role of the
North Star project, in which the United States is considering participation with the Soviet Union, a brief review of the trade arrangement
known as the joint venture is presented. It is reiterated here that
little documentation is available on Soviet joint ventures since the
concept of this type of arrangement is a recent phenomenon. The
information in this chapter is based on available literature.

DEFINITION OF A JOINT VENTURE

There are vast differences in the functioning of the Soviet planned
economy and the U.S. free-market economy. Since the phrase "joint
venture" is used in the context of both markets, it is necessary to
clarify some of the details of joint venture concepts.

Joint venture, or joint adventure, has a wide variety of meanings
in American business literature. It is usually defined as

> . . . a form of temporary partnership organized to carry
> out a single or isolated business enterprise for profit, and
> usually, although not necessarily, of short duration. It is
> an association of persons who combine their property,
> money, efforts, skill, and knowledge for the purpose of
> carrying out a single business operation for profit.[1]

Most narrowly defined, a joint venture involves equity interest,
that is, an investment by a country, either domestic or foreign,
involving joint ownership of the business. Some writers specify a
majority or minority joint venture, depending on whether the company

in question owns more or less than 50 percent interest in the enterprise. However, a joint venture may comprise any type of single
business deal or undertaking.[2] A joint venture differs from a partner-
ship in a legal sense. A joint venturer is not an agent of his co-
venturers and does not necessarily have the authority to bind them;
management and operation of the venture are specified by contract;
and a joint venturer can sue his co-venturer in a court of law if a
dispute arises between them.[3]

Prior to World War II, foreign investments by American business
were almost exclusively majority joint ventures; that is, the American
corporation owned a majority interest so as to assure control over
the investment. Political developments after the war brought pressure
for foreign control of local businesses. Colonial territories were
granted independence, resulting in a rise of nationalism. In some
cases, concern about economic exploitation and frustration over the
slow pace of economic progress brought restrictions on the operations of foreign-owned corporations or outright nationalization.
Independent developing countries began experiencing popular pressures
for national control over their natural resources. Even developed
countries began applying restrictive measures against corporations
controlled by foreigners, such as tariffs, quotas, and currency
limitations, in order to conserve scarce foreign exchange and to
balance international accounts. To overcome the various political,
legal, psychological, and economic barriers, foreign investors
increasingly turned to joint ventures to permit them to operate more
efficiently in foreign markets.

Originally, foreign corporations invested in a venture and,
occasionally, gave a degree of minority equity to local nationals.
Today, the situation is reversed. A Columbia University research
study found that, since 1961, "the joint international business venture
has become the predominant form of foreign investment in developing
countries."[4]

Various forms of equity joint ventures have been developed.
These include ownership agreements between private corporations,
between host governments and private parties, and between consortia
of foreign parties and host governments supported by loans from
national and international development aid associations. Joint
ventures have two major advantages. First, the risk of nationalization
is minimized. Second, the financing potential is increased. The
capital required for investment is large, particularly in energy
projects, and the financing potential is maximized in a joint venture
arrangement.

Joint ventures also have involved nonequity arrangements in
which risk and rewards are shared as set forth in a contractual

agreement, but equity ownership remains with the host country. An example of this type of arrangement is the agreement signed by the National Iranian Oil Company (NIOC) with five Western firms. The contract permits the Western firms to explore and develop oil production in South Iran. The Western companies supply funds and conduct operations. Refunds of expenses are dependent on discovery of oil, and the Western firms are guaranteed purchases of definite quantities of oil at favorable prices.[5]

Foreign investment in Communist countries has been limited because of the restriction on the private ownership of property. However, in commercial relations with East European countries, some Communist nations have modified their laws regarding private ownership of property to permit minority joint ventures. In July 1967, the Yugoslav foreign investment law was enacted. It was modified in 1971 to encourage foreign investment. Details of the law are[6]

1967 law

1. Foreign equity investment in Yugoslavia was permitted.
2. Maximum foreign equity of 49 percent was allowed.
3. Profit split was to be based on equity.
4. Profit repatriation was guaranteed, but 20 percent of profits must be reinvested in Yugoslavia.

1971 modification

1. Profit reinvestment requirement was eliminated.
2. Control must remain in Yugoslav hands.
3. Maximum tax was set at 35 percent, but taxes could be reduced if profits were reinvested in Yugoslavia.

In 1970-71, both Hungary and Romania passed laws permitting up to 49 percent equity by foreign investors.[7] Foreign investment has remained far below East European expectations in spite of the legislative changes. However, another form of business cooperation, the co-production joint venture, has begun to win favor with Western investors. In this arrangement, a part of a product may be manufactured in one nation and then shipped to a second nation, where it is assembled into a finished product. A contract sets forth the obligations, costs, and profit split between the participants. This arrangement avoids the ownership problem by dividing the production-marketing functions through a contractual agreement, in which responsibilities and benefits are carefully spelled out. The co-production joint venture has advantages for the East, because it avoids the burden of debt repayment, allows the East access to continuous updating with Western technology, and permits access to Western management and marketing techniques. It benefits the West through low-cost raw materials and labor, and it permits access to Eastern markets which would otherwise remain closed.

Soviet law does not permit private ownership of the "means of production." Therefore, the Soviets have proposed another type of joint venture. The foreign investor is invited to construct a plant or develop a raw material resource in the Soviet Union and is promised repayment in the form of output from the investment. The return would include interest and a reasonable profit.

As pointed out, the term "joint venture" can have many meanings. Most definitions, except those used by the Soviets, include equity rights. Introduction of a new term for the Soviet proposal, however, would introduce even more confusion. Therefore, joint venture will be used in this paper as the Soviets use it, that is, an investment without ownership title. Similarly, investment will refer to the construction of a production facility or development of a raw material resource, but will not automatically imply ownership as it does in Western usage.

An important factor included in the concept of Soviet retention of full ownership is that any debt incurred is a Soviet responsibility and not that of the participating foreign corporations. The Soviets would be required to repay the debt even if a project failed to function as planned.

THE ROLE OF LICENSING IN JOINT VENTURES

The joint venture is not a new or novel arrangement for the Western business community, for the original practice dates back many years. East India companies were established in the seventeenth and eighteenth centuries in England, Holland, France, Denmark, Scotland, Spain, Austria, and Sweden to exploit trade with India and the Far East. The British East India Company was incorporated by royal charter on December 31, 1600.[8] Most of these companies penetrated international markets simply by exporting their products to them. Trade practice gradually evolved into licensing and joint venture arrangements.

Licensing involves the granting of a patent, trademark, or know-how rights in accordance with the terms of a contract based on the products or technology involved.[9] It can be a cheap, profitable method of market entry, for it provides entry without risk of investment. For emerging industries in developing countries, technical

assistance through licensing can provide an opportunity for entry
into a gradually developing market, which would not otherwise justify
a large investment. Furthermore, the markets of developing economies
are often closed to imports by tariffs or other barriers.

Licensing poses the following disadvantages and hazards, however, as a marketing technique:

1. Control over the licensee's manufacturing and marketing operations
   is rarely satisfactory.
2. Licensing is probably the least profitable method of exploiting a
   foreign market.
3. Every licensee is a potential competitor.[10]

Today, licensing is employed in U.S. exchanges with the Soviet
Union and East European countries and is normally confined to specific, limited contracts, thus minimizing standard risks. An example
is a U.S. petrochemical firm with sophisticated technical expertise
which recently sold packages of technology to four East European
countries after lengthy negotiations. A West European construction
firm managed the entire licensing transaction; that is, it constructed
the plant, supplied equipment, and arranged the financing. The American firm supplied the technology and services, in the form of advice,
during plant construction. The contract provided for payments in cash
according to a specified schedule, which was completed before the
plant was opened.[11]

Licensing is not discussed further in this study because of its
limited applicability to foreign investment. It has been mentioned
only because licensing arrangements in the past have involved or led
into joint ventures.

THE IMPORTANCE OF CONTROL

Following World War II, the developing countries of the world
insisted on legal and political sovereignty over their natural
resources; however, they lacked investment capital, technical skill,
and managerial know-how. Prior to the war, foreign investors had
been accustomed to having complete control over their investments.
In the political environment of the postwar era, such control was
considered economic exploitation. The conflicting realities forced
an accommodation in which each party compromised some of its
prerogatives through the structure of joint ventures.

In structuring joint ventures, the problem of control is vitally
important, but it is extremely difficult to work out concrete terms

for an agreement that would provide for effective control of a business. A special study of the problem of control was made by Jean-Pierre Beguin in connection with a Columbia University research project on joint ventures. The study showed that equity control does not necessarily assure real control over the undertaking.[12] It is not so much the formal arrangement as the way the terms of the agreement are carried out that provides control over the enterprise.

According to Beguin, voting arrangements can be structured to provide control in a minority equity investment. For example, in the multinational joint venture Fria in Guinea, control resides in the French aluminum company Pechiney through a device of "decuple-vote shares"; each Pechiney share is entitled to 10 votes and all other shares to 1 vote.[13]

The study shows that some corporations insist on management control as a matter of policy and gives the following examples:

1. The German steel company Mannesmann insists on control as the only sure way to apply its own technological skills.
2. International Business Machines (IBM) insists on control to safeguard complex technology. The corporation employs a highly developed specialization process, and elements manufactured by different participating units must meet extremely precise technical and quality standards.
3. Merck and Company insists on control to protect its patents and know-how.[14]

The study contends that the nature of the product a corporation manufactures and the business philosophy it espouses determine its policy toward joint ventures.

Beguin's study also reveals that although the majority equity arrangement can provide advantages in exercising the control function and can improve operations and efficiency of an enterprise, host government actions in establishing legal, financial, and social requirements can, and often do, negate these advantages. In addition, other factors besides voting control contribute to the success of an undertaking, such as compatible business partners, the legal and financial structure, extensive employment of nationals, and local ownership and support. Where feasible, public offerings of shares promote this support. The general political and psychological attitude that prevails in the international business community today is forcing the foreign investor to accept minority equity and contractual joint ventures despite some of the disadvantages inherent in these arrangements. The study declares

. . . foreign investors (notably in the oil industry) stress
the fact that, in their experience, cooperation between
partners has not been any more difficult to achieve in contractual joint ventures than in equity joint ventures. In the
basic agreement, the partners have every latitude to provide one or several executives with extensive powers and
with the means of using these powers. . . . In principle,
the management in a contractual joint venture cannot exceed
the precise budgetary limits that have been agreed to. . . .
In practice, however, the budget may be planned so as to
leave a sufficient margin of financial freedom to the senior
executive. . . . It can therefore be concluded that the
flexibility inherent in a contractual joint venture is far
from being necessarily an obstacle to the cooperation between
the partners and to the efficiency of management.[15]

JOINT VENTURES IN DEVELOPING COUNTRIES

In 1968, Lawrence G. Franko, as a dissertation proposal at
Columbia University, undertook a study to determine the ultimate
fate of joint ventures.[16] He reported that during 1964 a group of 170
U.S. multinational firms had entered into approximately 1,100 joint
ventures in countries where the U.S. partner could have legally
chosen wholly owned subsidiaries. Over a period of six years,
ownership changes occurred in 314, or approximately one-third, of
the 1,100 firms as follows: 182 became wholly owned subsidiaries;
84 were sold; 46 U.S. firms became the majority owner (control
over 50 percent equity); and 2 foreign firms took control (assumed
over 50 percent equity).[17]

According to the study, an important reason for the changes in
the joint ventures was given as disagreement over production,
marketing, and research and development strategy. Corporations
following a strategy of product-market concentration, that is,
serving one product market, found joint venture partners to be a
hindrance. Corporations with decentralized decision making, however, which constantly diversified their interests abroad by introducing new products into foreign markets, successfully employed
and had a high tolerance for joint ventures.[18]

In his study, Franko surmises that joint venture experience in
the Western business environment indicates that the strongest appeal
of the joint venture arrangement is the reduction of political risk and
diversification of economic risk. The local partner guards against

government encroachment or outright appropriation. In developing countries, joint ventures are popular because local nationals are given the opportunity to share in profits and management. The foreign partner introduces technology and capital, which build up local industry, supply services, and expand exports, thereby promoting development of the economy.

Franko also claims that the primary objection to minority joint ventures (less than 50 percent equity) raised by U.S. executives was that it delayed decision making and curbed freedom of action in production and marketing decisions.[19] The main issue centered largely on the matter of control, which was exercised on the basis of equity voting rights. This function is particularly important in the areas of dividend policy, major expansion, borrowing money, selection of management, acquisitions, and dissolution.

A different view is suggested by Kindelberger, of the Massachusetts Institute of Technology, who believes that relationships between the participants in a business venture are more important than the voting rights in the formal equity arrangement. He says

. . . Control is not an either-or proposition, but a question of infinite degrees of divisibility, depending upon the nature of the decision-making process and the division of authority between the head office and the foreign unit. This control may cover any or all of a variety of separate functions—hiring and firing, investment programming, research and development, pricing, dividend remittances, marketing, and so on. A company can control all phases of a subsidiary's operations with merely 25 percent of the equity, on the one hand, or it may passively receive dividends without interfering in any of the affairs of its 100-percent-owned foreign operation. In the latter case, it is in effect merely a portfolio owner.[20]

EVOLUTION OF THE CONTRACTUAL JOINT VENTURE IN IRAN

In 1956, a Columbia University research project under the direction of Wolfgang G. Friedmann undertook to study joint international business ventures between developed and developing countries. The data produced were published in 1971 in a book entitled Joint International Business Ventures in Developing Countries, which describes the evolution of the contractual joint venture.[21]

According to Friedmann, one of the first developing nations to invite foreign investment was Iran. In 1957, Iran passed a petroleum law which provided for two kinds of foreign investment: the traditional equity joint venture and the contractual joint venture. The Iranian government favored contractual joint ventures, because this arrangement permitted greater control over the country's natural resources. After negotiating for four years, NIOC signed a joint venture agreement on January 17, 1965, with three countries for the exploration and production of offshore oil in the Persian Gulf.

The agreement established a legal structure of two parties as follows:[22]

First party (NIOC)—50 percent
Second party (foreign interests—50 percent
   One-third—Phillips (American Oil Corporation)
   One-third—Assienda Generale Italiana Petroli (AGIP)
   One-third—Oil and Natural Gas Commission of India (ONGC)

The agreement stipulated that the second party was in charge of all exploratory operations and was responsible for the entire burden of expenses. After commencement of commercial production, expenses were to be shared equally, and the production was owned directly by the partners. Management was in the hands of a joint operating corporation, the Iranian Marine International Oil Company. During exploratory operations, it acted as the agent of the second party. After production commenced, it acted as an agent of both parties, with the chairman of the board nominated by the first party and the vice chairman-managing director nominated by the second party.

The agreement stipulation that production was owned jointly by the partners provided the equity guarantees which made the agreement acceptable to the foreign partners. For the American Oil Corporation, this clause was particularly attractive. When direct ownership can be established, U.S. law permits deduction of such expenses as research and exploration costs from taxable income in the year of occurrence. In addition, a depletion allowance is permitted, which represents a 22 percent deduction from taxable profit.[23]

On March 3, 1969, NIOC signed another agreement with five West European firms for exploration and development of oil production in South Iran.[24] The agreement represented the achievement of Iran's goal—the contractual joint venture. By the 1965 agreement, the Iranian government owned 50 percent of the oil production. By the 1969 agreement, it owned 100 percent of the production.

The 1969 agreement is a service contract, whereby the foreign firms agree to undertake oil exploration and production operations in return for guaranteed rights to purchase oil, if found, at favorable prices. The 1969 agreement has disadvantages for U.S. corporations over the 1965 type of agreement, because, under U.S. tax law, absence of ownership rights precludes deduction of expenses and the depletion allowance. In essence, the 1969 agreement is a civil law service contract.

JOINT VENTURES IN YUGOSLAVIA

A Communist country that is open to joint business ventures is Yugoslavia. On January 31, 1946, it adopted a constitution that was based on the principles of social ownership and worker self-management of the economy. By law, foreign capital investment was excluded from the republic.

In 1965, however, basic economic reforms, which decentralized the economy, were promulgated. Direct state control over enterprises was reduced. Autonomy and competition among enterprises were encouraged. The need to increase foreign trade earnings was recognized, and planners began to seek the means of stimulating the development of Yugoslav industry. It was decided that foreign investment and technical assistance would be welcome, provided it did not interfere with the basic principles of the Yugoslav constitution.

Principles for collaboration with foreign enterprises and modification of relevant laws were drafted by the Yugoslav government, and a joint colloquium was held at Belgrade in 1967 to discuss the legal, economic, and managerial problems of joint ventures prior to enacting the proposed changes. Wolfgang G. Friedmann, of Columbia University Law School's International Legal Studies Program, and Leo Mates, of the Institute of International Politics and Economy, Belgrade, who served as co-chairmen of the colloquium, provide a summary of the discussion in a book published in 1968.[25]

According to Friedmann and Mates, it was agreed that majority joint equity, where the foreigner owned majority equity interest, was illegal. "At least in the near future, the Yugoslav participants thought there was no possibility to apply this form of joint venture in Yugoslavia,"[26] the summary reported, quoting Article VI of the Yugoslav constitution, which states:

> The social and economic system of Yugoslavia is based on freely associated work with socially owned means of labor and self-management of the working people in the production and distribution of social product in the working organization and social community.[27]

In their book, Friedmann and Mates declare that many of the participants in the colloquium, both Yugoslav and foreign, believed that all legitimate interests of both foreign investor and domestic partner could be reconciled and protected in a contractual joint venture. It was agreed that a very elaborate contract could be written to satisfy all interests, but it was not certain that such a contract would permit smooth operation of a business. The most difficult problem would be the position and function of the manager, because, under the worker self-management concept, the workers exercise some of the functions of management—in a sense, they are self-employed. It was felt that if there was goodwill on the part of domestic enterprise and foreign investor, an accommodation could be reached.[28]

Later on, in 1967 and 1971, legislative changes were made by the Yugoslav government legalizing minority joint ventures (up to 49 percent foreign ownership). The changes, however, still do not permit real proprietary interest by foreigners, nor do they allow representation on the ultimate decision-making bodies of enterprises. Unfortunately, the 1971 amendments did not generate the increase in foreign investment that the Yugoslav government expected.[29]

JOINT VENTURES IN THE USSR

Prior to the Communist seizure of power in October 1917, Russia was a respectable industrial power, according to Anthony Sutton. The economy included several hundred medium-to-large manufacturing enterprises, with airplanes and automobiles of indigenous Russian design being produced in quantity. There were obvious signs of Russian technology in chemicals, turbines, and railroad equipment. The manufacturing complex was supported by numerous self-contained mining enterprises in the Urals and DonBas. The International Harvester plant at Omsk was the largest in the company's worldwide network. (This structure was substantially intact after the Communist revolution.)[30]

When Lenin came to power in 1917, he believed that the economic system devised by capitalism for the allocation of resources and distribution of production was characterized by greed and exploitation. His goal was to reform Russian* society and to free it from the

---

*"Russian" is used loosely in this study to mean "Soviet" because the word appeared in this form in research literature. Technically, the Russian state ceased to exist when the revolution was successful.

corruptive influence of private ownership and the contaminating environment of the capitalistic system. Lenin nationalized heavy industry, the banking system, foreign and domestic trade, all handicraft, and other small industries. Money was inflated to zero value, and an effort was made to wipe out all elements of the free-market and private economic activity. Lenin's program was based on barter, the direct exchange of commodities between town and countryside without market relations or trade, in the hope of achieving his goal of socialism immediately.[31]

Widespread economic disorganization, however, soon placed the Soviet economy on the verge of bankruptcy. Lenin retreated and admitted that he had made a mistake, explaining that War Communism (his economic decrees) was necessitated by the pressure of foreign intervention and conceding that the economic program had not been carefully worked out. In the speech introducing his New Economic Policy (NEP) to the Tenth Party Congress, he ascribed War Communism to the "dreamers" who supposed that it would be possible in three years to transform the economic base of the Soviet order.[32] To prevent the impending bankruptcy of the Soviet economy, NEP made the following substantial compromises with ideology:

1. Private industrial production and private trade were legalized.
2. Free-market incentives were substituted for coercion.
3. Foreign capitalists were encouraged to invest in the Soviet Union.
4. Free contract among enterprises replaced the centralized allocation of raw materials and equipment.
5. Tight control was retained over the "commanding heights" of the economy. Control of key industries, transportation, communications, banks, and foreign trade remained in the hands of the state.[33]

Lenin feared foreign contacts both from an ideological (corruptive influence) and a national security (foreign intervention) standpoint. He continued to search for some form of ad hoc economic collaboration that would specifically benefit the industrial sector of the Soviet economy but would remain at all times under the effective control of the Soviet government.[34]

The NEP denationalized certain economic activities and restored a measure of free enterprise to both foreign and domestic capitalists. Internally, controls were relaxed in the area of retailing, wholesaling, and small industries employing fewer than twenty persons. The "commanding heights," including iron and steel, electrical equipment, transportation, and foreign trade, were grouped into trusts and syndicates. Foreign capital was invited into these units through concessions and mixed joint-stock companies, both with and without domestic-private and state participation. The concession, in its varying forms, was the most significant vehicle for the transfer of foreign technology.[35]

The concession was, in effect, a joint venture, and was suggested in December 1917 at the first all Russian Congress of the Councils of the National Economy. Although subsequent negotiations with foreign capitalists were temporarily halted by the allied intervention and the civil war, the law of August 21, 1923, established the legal structure for the conduct of negotiations and the transfer of Russian property to foreign enterprises.[36] Under this law, concessions were divided into three categories:

1. The "pure" concession (or Type I) was an agreement between the USSR and a foreign enterprise, whereby the foreign firm was permitted to exploit an opportunity within the USSR under the legal doctrine of usufruct, that is, without acquiring property rights. Royalty payments to the USSR were an essential part of the agreement. In all cases, the foreign enterprise was required both to invest a stipulated amount of capital and to introduce the latest in Western technology and equipment. The Type I concession closely resembled the current contract joint venture.

2. The "mixed" company concession (or Type II) utilized a corporation in which Soviet and foreign participation were on an equal basis, at first 50-50 but later 51-49 percent equity. A Soviet chairman of the board had the deciding vote in cases of dispute. Normally, the Soviets provided the investment opportunity and location, and the foreign company provided capital, technology, and skills. Profits were split in accordance with equity.

3. The Soviets designated the technical assistance contract as a Type III concession. In essence, Type III was a reverse technical concession in that the Soviets were making payments to exploit foreign technology. Type III concessions resembled the current licensing agreements.

At the beginning of the NEP, the emphasis in the Soviet Union was on concessions to Western entrepreneurs. In the middle and end of the decade of the 1930s, the concession was replaced by technical assistance contracts and "turnkey" contracts, that is, the importation of complete plants and equipment. In size, concessions ranged from the gigantic Lena Goldfields of the United Kingdom, operating 13 separate industrial complexes and valued (after Soviet expropriation) at over $89 million, to small factories manufacturing pencils (the Hammer concession) or typewriter ribbons (the Alftan concession).[37] Highlights of the concessions agreements are summarized in Table 2.1.

TABLE 2.1

Soviet New Economic Policy Concessions of 1921

Type I: Pure Concession—Foreign Firm Develops and Exploits Opportunity:

1. No property rights
2. Royalty payments to the USSR
3. Foreign firm required to invest stipulated amount of capital and required to introduce latest Western technology
4. Firm contract for 20 to 30 years
5. Profit and capital repatriation warranty

Type II: Mixed Company—Mixed Soviet-Foreign Participation (50:50, and later, 51:49 ownership):

1. Soviet chairman has deciding vote
2. Foreign firm provides capital and technology
3. Soviets provide investment opportunity and specify location
4. Labor partly imported
5. Profits split according to equity

Type III: Technical Assistance Contract:

1. Soviets agree to pay foreigners for technology (patents, design, know-how)

Source: Anthony C. Sutton, Western Technology and Soviet Economic Development, 1917 to 1930 (Stanford, Calif.: Hoover Institution, 1968), pp. 6-11.

Armand Hammer, who was granted the first foreign concession in the Soviet Union in 1921, described his pioneering efforts to build trade between the USSR and the United States in an article published in the American Review of East-West Trade. Hammer stated that he visited Russia immediately after completing medical school in 1921. Moved by the poverty and suffering of the Russian people, he offered to use his million-dollar estate to buy food and ship it to the Soviet Union. The Soviets were impressed by his offer and arranged a meeting for him with Lenin. In the discussion that followed, the exploitation of asbestos and the manufacture of pencils were discussed, and concessions were awarded to him.[38]

Anthony Sutton confirms that a Type I concession was granted to the Allied Chemical and Dye Corporation of the United States, whose subsidiary, the Allied American Corporation, owned by the Hammers, had been operating under license in the USSR since 1918. The concession was to restore and operate the Alapaievsky asbestos mines.[39]

After Hammer returned to the United States, he launched a "grain for furs" campaign, which sought to encourage American business to invest in Russia. He spoke with Henry Ford and officials of 38 companies in every field of manufacture to inform them of the opportunity for investment in the USSR. He organized and served as secretary to the Allied American Corporation and also managed the Alapaievsky asbestos concession.[40]

The NEP initiated by Lenin proved to be a success. By 1928, the Soviet economy had recovered, and prewar production levels were attained. As Stalin grasped the reins of leadership, the concern about foreign intervention turned into a paranoic fear about the security of the Soviet Union. Stalin realized that, through the concession agreements, domestic and foreign capitalists (mostly German because of the Rapallo Treaty and associated protocols) possessed a stranglehold over the Soviet economy. He realized that access to Western technology was needed, but felt that control arrangements were unsatisfactory. Through his efforts, emphasis shifted to technical assistance contracts whereby "turnkey" plants were constructed and management was placed under control of reliable nationals. Where Type I and Type II concessions were required, contracts were signed with tightly specified conditions and for short-term periods.[41]

Where long-term contracts existed, Stalin employed several tactics which made the concession unprofitable for the businessman and led to expropriation. Among measures used were labor strikes, raising taxes, charges of economic espionage, and bribery of Soviet officials. In 1928, the "Shakhta Affair" was the first of many show trials. Five German engineers were charged with "a counterrevolutionary plot to destroy and disorganize the coal industry"[42] and were accused of having links with former mine owners and the Polish counterespionage service. It was said that they had started fires, created explosions, wrecked coal-cutting machines, broken down shafts, and generally created mayhem in the mines. In summary, they were accused of sabotaging "socialist construction." U.S. State Department archives contain a number of foreign government reports on the affair, and their consensus is that the real reason for the arrests was the dominant place achieved by the Germans in Russian industry. The Germans had become too powerful and threatened the control of the party. The move was a broad-based attempt by the Communist party to reassert its control over industry, and measures were taken against the united front of specialists, old-time Russian engineers, trade unions, many of the workers, and some of the "red" plant directors, who were suspected of not being loyal to the party.[43]

By the mid-1930s, Stalin's paranoid fear of foreign intervention caused the concession almost to disappear from use. Stalin's strategy for Soviet national security was based on the achievement of economic autarky, that is, complete independence and isolation of the Soviet economy so as to free it from subversion or dependence on the outside world.[44] This policy was to remain in effect until after World War II.

Review of Dissertation Abstracts

A search of <u>Dissertation Abstracts</u> from 1969 to the present failed to provide information of major significance to this study.[45] A recent dissertation, however, "Export Potential of the United States Tool and Die Industry to the Soviet Union through the Sale of Turnkey Plants," by Lyn Floyd Wheeler,[46] does have marginal significance to this study. Wheeler analyzes the U.S. tool and die industry to determine its ability to provide technology sought by the Soviet Union and its capability for selling this technology in the form of complete turnkey facilities. The objective was to determine whether small business could significantly increase exports to the Soviet Union. Wheeler's research is limited to a relatively small project sold under conditions of normal trade. Barter was not involved. Government financing did not appreciably affect the sales because of size and time factors. In the present study, the barter-trade concept is evaluated in the North Star project, which involves expenditure of large amounts of money and covers a time span of 25 years. Also, the role of government financing assistance in promoting long-term barter trade is analyzed.

SUMMARY

In this chapter the joint venture arrangement was defined, and the evolution of the current joint venture concept was traced from its origins through the years. Foreign direct investment practice has evolved from an insistence on majority equity to an acceptance of minority equity or even service contract arrangements. Iran, Yugoslavia, and Russia have all experimented with different types of joint ventures.

It has been demonstrated that control is an important function in a joint venture project, and adequate provisions for the effective exercise of control are essential for the efficient operation of such an undertaking. It is not necessary to possess majority equity, however, to exercise effective control. In fact, cooperation between partners in a contractual joint venture can achieve efficient management and offers certain advantages in reducing political risk and diversifying economic risk. A contractual service contract has been signed and implemented in Iran. It sets forth exploration and development obligations and grants rights to purchase the petroleum produced at concessionary rates.

Western business operated in the Soviet Union after 1921 on the basis of concessions. Political factors played a role in the granting of concessions and, eventually, caused them to be phased out. This political factor is present in today's environment, but the political environment, both within the Soviet Union and worldwide, has changed since that earlier period.

NOTES

1. Len Young Smith and G. Gale Robertson, Business Law: Uniform Commercial Code, 3rd ed. (St. Paul, Minn.: West Publishing Co., 1971), pp. 736-37.

2. Ibid., pp. 742-43.

3. See generally "Joint Adventures," 48 Corpus Juris Secundum 801 (1947).

4. Wolfgang G. Friedmann and Jean-Pierre Beguin, Joint International Business Ventures in Developing Countries (New York: Columbia University Press, 1971), p. vi.

5. Ibid., pp. 52-53.

6. Timothy P. Neumann, "Joint Ventures in Yugoslavia: 1971 Amendments to Foreign Investment Laws," New York University Journal of International Law and Politics 6 (summer 1973): 271-96.

7. Hungary, Magyar Koz'ony Number 67 [Hungarian Law Number 67] Budapest, August 7, 1970; and Nicolae Ceausescu, Law on Foreign Trade and Economic and Technico-Scientific Cooperation Activities in the Socialist Republic of Romania (Bucharest: Chamber of Commerce of the Socialist Republic of Romania, 1971).

8. The New Encyclopaedia Britannica, 15th ed., Micropaedia, Vol. 3, Ready Reference (1973), p. 762.

9. Vincent R. Travaglini, "Licensing, Joint Ventures, and Technology Transfer," International Commerce, July 28, 1969, p. 2.

10. Ibid.

11. Robert S. Kretschmar, Jr., and Robin Foor, The Potential for Joint Ventures in Eastern Europe (New York: Praeger Publishers, 1972), pp. 8-9.

12. Jean-Pierre Beguin, "The Control of Joint Ventures," in Joint International Business Ventures, ed. Friedmann and Beguin, pp. 364-418.

13. Ibid., p. 367.

14. Ibid., pp. 367-68.

15. Ibid., p. 418.

16. Lawrence G. Franko, Joint Venture Survival in Multinational Corporations (New York: Praeger Publishers, 1971).

17. Lawrence G. Franko, "Joint Venture Divorce in the Multinational Company," Columbia Journal of World Business (May-June 1971): 13-22.

18. Ibid.

19. Ibid.

20. C. P. Kindelberger, International Economics, 3rd ed. (Homewood, Ill.: Richard D. Irwin, Inc., 1963), p. 404.

21. Friedmann and Beguin, op. cit.

22. Ibid., pp. 38-54.

23. Ibid., p. 42.

24. Ibid., pp. 52-54.

25. Wolfgang G. Friedmann and Leo Mates, Joint Business Ventures of Yugoslav Enterprises and Foreign Firms (Belgrade, 1968), pp. 13-40.

26. Ibid., p. 23.

27. Ibid., p. 22.

28. Ibid., pp. 23-28.

29. Neumann, op. cit., p. 296.

30. Anthony C. Sutton, Western Technology and Soviet Economic Development, 1917 to 1930 (Stanford, Calif.: Hoover Institution, 1968), pp. 344, 310-11, 5259-60, 5310-11.

31. Yevsei Liberman, "The Soviet Economic Reform," Foreign Affairs 46 (October 1967): 53.

32. John P. Hardt, "Soviet Economic Development and Policy Alternatives," in The Development of the Soviet Economy, ed. V. G. Treml (New York: Praeger Publishers, 1968), pp. 5-6, quoting E. H. Carr, History of Soviet Russia: The Bolshevik Revolution (London: Macmillan & Co. Ltd., 1952), p. 275.

33. Harry Schwartz, The Soviet Economy (Philadelphia: Lippincott, 1965), pp. 12-13.

34. Leon Herman, "The Promise of Economic Self-Sufficiency under Soviet Socialism," in The Development of the Soviet Economy, ed. Vladimir G. Treml (New York: Praeger Publishers, 1968), pp. 215-16.

35. Sutton, op. cit., p. 5.

36. Ibid., pp. 6-9.

37. Ibid., pp. 6-10.

38. Armand Hammer, "American Entrepreneur—First Foreign Concessionaire in the Soviet Union," American Review of East-West Trade (March-April 1970): 14-22.

39. Sutton, op. cit., pp. 108-09.

40. Ibid., p. 109.

41. Ibid., pp. 348-49.

42. Ibid., pp. 325-26.

43. Ibid., p. 325, citing U.S. Department of State, Decimal File 316.6221/13, 316/6221/25, and 316.6221/28.

44. Joseph S. Berliner, Soviet Economic Aid (New York: Frederick A. Praeger, Inc., 1958), pp. 80-82; and U.S. Congress, Joint Economic Committee, Subcommittee on Economic Statistics, Comparisons of the U.S. and Soviet Economies, pt. 2, 86th Cong., 1st sess., 1959, pp. 414-16.

45. Dissertation Abstracts International. A: The Humanities and Social Sciences (Ann Arbor: Xerox University Microfilms, vol. 30, no. 1, July 1969, to the most recent volume available, vol. 34, no. 8, February 1974).

46. Lyn Floyd Wheeler, "Export Potential of the U.S. Tool and Die Industry to the Soviet Union through the Sale of Turnkey Plants," D.B.A. dissertation, The George Washington University, 1974.

3

THE NATURAL GAS
SUPPLY PROBLEM
IN THE UNITED STATES

Employment of a joint venture arrangement to finance the North Star project requires major departures in the manner of doing business both in the United States and in the Soviet Union. Further movement along the path of detente is required, including enactment of legislation to overcome restrictions included in the Trade Act and Export-Import Bank Amendments of 1974. The seriousness of these actions suggests that the North Star project must rest on unassailable foundations so as to be economically and financially feasible. In order to provide the impetus to make the necessary changes, there must be serious need by both the Soviet and the U.S. economies, and North Star must contribute toward resolution of these needs. Discussed in this chapter will be the U.S. need—a shortage of natural gas.

The North Star project is an outgrowth of the U.S. natural gas supply problem. Is there a shortage of gas, and will it continue in the future? Given that a shortage does exist, is there a better alternative than importing gas from Siberia? Would the importing of gas from Siberia be cost competitive in the U.S. market? These questions must be answered affirmatively to justify embarking on the North Star project.

Before the 1973 Arab-Israeli war accelerated the realization that imported energy could not be relied upon to satisfy U.S. needs, voices in business, government, and academia were warning of an impending energy crisis.

In 1970, Lawrence Rocks and Richard P. Runyon attempted to publish their book, The Energy Crisis, and were turned down by five of the largest names in publishing. The publishers said they were not aware that an energy crisis existed, and they did not believe that a story about it would be of interest to the public.[1] The book was finally published in 1972.

The National Association of Regulatory Utility Commissioners adopted a resolution on July 17, 1969, acknowledging the existence of and the need to solve the gas supply problem. A Federal Power Commission (FPC) staff briefing, published on April 15, 1971, stated flatly that evidence submitted to the commission "confirms beyond any doubt, if indeed there is any remaining doubt, that a serious gas supply shortage does in fact exist throughout the nation's gas supply areas."[2]

A contributing cause of the natural gas supply shortage appears to be the pricing policy which, in effect, froze gas prices at the wellhead at the 1955 level.[3] Data show that while wellhead prices remained relatively constant, drilling costs escalated, which resulted in a 40 percent decrease in wildcat drilling between 1956 and 1970. Prior to 1968, supply was not a problem, because the annual discovery of new reserves exceeded the annual consumption of gas, as reflected in Table 3.1. However, the reserves-to-production ratio continued to decrease and declined from 16.4 in 1966 to 11.9 in 1970.[4] On December 16, 1968, the American Gas Association (AGA), representing 300 companies which service 92 percent of the gas customers in the United States, warned the FPC that gas distributors were having difficulty in contracting for increases in long-term gas supplies. The FPC was requested to provide additional economic incentives for exploration and development.[5]

The gas supply situation, as reflected in the ratio of total reserves to production, grew progressively worse, as indicated in Table 3.1.

In June 1969, some 10 distributor executives, servicing approximately 40 percent of all gas customers, reported to the FPC that they could experience actual, though isolated, shortages in the winter of 1969-70. In the winter of 1969-70, one company had to curtail deliveries to major industrial plants for six days during a peak January cold spell. Some gas companies in the industrial Midwest rejected accounts which they had sought for years.[6]

Within the Washington metropolitan area (Washington, D.C., Maryland, and Virginia), the Washington Gas Light Company, after obtaining regulatory approval, refused to accept new commercial customers in October 1971 and new residential subscribers in March 1972. The company services 550,000 customers, and gas supply delivery is 4 percent below that contracted for under suppliers' agreements. The price of gas supply is 57 cents for each 1,000 cubic feet. The company recently contracted to buy gas from the Green Springs, Ohio, synthetic plant (gas manufactured from naphtha derivatives) at the estimated price of $2 per 1,000 cubic feet. The gas was scheduled to begin flowing in January 1974. The Washington Gas Light Company is currently constructing its own synthetic gas plant. It has also contracted to purchase liquefied natural gas (LNG) from Algeria via the Columbia Gas System. LNG is scheduled to be

TABLE 3.1

Status of Consumption and Reserves of U.S.
Natural Gas (in trillion cubic feet)

Year	Consumption	New Reserves Added	Total Reserves to Production Ratio
1966	17.9	19.2	16.4
1967	18.8	21.1	15.8
1968	19.9	12.0	14.6
1969	21.3	8.3	13.1
1970	22.6	11.1	11.9*

*Excluding Alaska.

Note: Actual data on consumption and new reserves were published only through 1970. John N. Nassikas reported at the Fourth International Conference on Liquefied Natural Gas, held in Algeria on June 24, 1974, that U.S. domestic production of natural gas reached 22.5 TCF in 1973, compared with 21.8 TCF produced in 1970. Therefore, production is slightly higher, but consumption has been held down by the refusal of suppliers to take on new customers.

Sources: Compiled from data contained in American Gas Association, Natural Gas Supply Problem: Background Report (Arlington, Va., 1972); and U.S. Federal Power Commission, National Gas Supply and Demand, 1971-90, Staff Report No. 2 (Washington, D.C.: Bureau of Natural Gas, February 1972).

delivered at Cove Point (Wilmington, Delaware) commencing in 1977. The cost is expected to be slightly more than $1.50 per 1,000 cubic feet.[7] In 1972, regulatory authorities authorized a purchase price adjustment clause for rate determination, which would permit the company to pass price increases directly to customers on a monthly repricing basis.

On February 23, 1971, the FPC authorized the establishment of natural gas survey advisory committees to

. . . accomplish the objectives of the Natural Gas Act, in providing for the ultimate consumer an adequate and reliable supply of natural gas at a reasonable price and the nation a vital resource base, the commission will direct the conduct of the survey through the members of the commission and its staff.[8]

The FPC staff, by an order dated December 21, 1971, was directed to undertake an independent analysis of the nation's proven natural gas reserves. The analysis was conducted through the combined efforts of the FPC staff, the United States Geological Survey of the Department of the Interior, the Office of Naval Petroleum and Oil Shale Reserves of the United States Navy, the Office of Management and Budget, the Bureau of the Census, and the regulatory and conservation agencies of the major gas-producing states. The results of the analysis were published in May 1973, and revised the following September, as the Natural Gas Reserves Study (NGRS). A comparison with the reserve data published by the AGA showed that the FPC study included 62 natural gas fields that met the criterion of having proven economically recoverable gas reserves but were not in the AGA data. Despite this omission, the total gas reserves estimated by the AGA were approximately 10 percent higher than estimated by the NGRS.[9]

Table 3.2 summarizes the results of the NGRS findings. The difference of 28.1 trillion cubic feet (TCF) in the gas-reserve estimates is due primarily to different estimates of associated and nonassociated gas reserves recorded for 6,358 entries and represents 10 percent of total reserves. Essentially, the government estimate of reserves was more conservative than that of the gas industry. Historically, the AGA reserves-to-production ratio is used to compute gas production data, but the NGRS used less optimistic figures. The NGRS states that

> . . . projections of future production of natural gas from
> proven reserves which have been based on AGA figures
> should still be considered reasonable. However, they may
> be optimistic and the natural gas available from this source
> in the future may be more limited than previously reported.[10]

In other words, the long, meticulous study by the FPC indicated that the gas industry was not conspiring to create a false fear to to increase the price. On the contrary, a real gas shortage existed, and the industry was overly optimistic in estimating its ability to supply natural gas in future years.

On May 15, 1973, the Subcommittee on Energy of the House Science and Astronautics Committee held hearings on the research and development programs of the National Science Foundation designed to meet the challenge confronting the nation on energy. The following excerpts from the testimony of H. Guyford Stever, director of the National Science Foundation, reflected the extensive effort and investment already expended or contemplated for the purpose of developing energy alternatives in order to provide options to the nation for the formulation of an energy strategy:

TABLE 3.2

Status of National Gas Reserves in the United States
(in trillion cubic feet)

	National Gas Reserves Study	American Gas Association	Difference
Nonassociated and associated gas:			
Reported fields	225.5	252.0	+26.5
"Omitted" fields	0.1	—	- 0.1
Dissolved gas	33.0	34.7	+28.1
Total*	258.6	286.7	+28.1

*Excludes gas in underground storage.

Source: U.S. Federal Power Commission, Natural Gas Reserves Study, National Gas Survey Staff Report, revised (Washington, D.C.: Federal Power Commission, September 1973), p. 1.

. . . the federal government will allocate an estimated
$771.8 million. . . . Private industry is also extensively
involved . . . between $1.1 and $1.2 billion will be spent
. . . during fiscal year 1974 on energy related R. & D. . . .
. . . The major alternative to fossil fuel energy for
at least the remainder of this century is fissionable nuclear
fuel . . . today the number of nuclear plants on line or under
construction equals the entire electrical generating capacity
of the United States in 1950 . . . fast breeder reactor . . .
can reach the commercial demonstration stage by 1980 . . .
by the 1990s the breeder reactor will begin to be a major
contributor to the production of energy in the United States.
. . .
. . . Federal government obligations for coal related
to R. & D. in fiscal 1974 are planned at a level of approximately $120 million. Private industry will spend over $40
million. . . . If these pilot plants prove to be successful
. . . We could see from 12 to 37 such plants in operation
by 1985. It is, however, important to realize that it will
be in the 1990s before coal gasification can substantially
supplant our natural gas supply. . . .
. . . natural gas ranks second only to oil . . . the
federal R. & D. program in this area will approximate $4

million. . . . These funds support the AEC's nuclear gas
stimulation program . . . an additional $100 million in
R. & D. will be invested by the gas utility and associated
industries . . . coal gasification . . . fuel cell technology
. . . and a sizeable portion in synthetic gas research and
development. . . .

    . . . I believe that we will have problems, serious
problems with energy on a short-term basis for years.[11]

Whether one looks at the current supply situation or the possible
research and development alternatives, the future is not bright in the
short term. The data presented indicate that a real shortage of
natural gas exists. How it came about is a matter for much argument.
It is generally conceded that a price held artificially low was a major
contributor in that it stimulated demand and promoted wasteful
usage.[12] In 1972, the FPC and regulatory agencies permitted prices
to increase on newly discovered deposits to reflect the increased
exploration costs. The decline in exploratory gas well drilling, which
commenced in 1956, was reversed in 1972. The trend has continued
upward in 1973, increasing by 50 percent.[13] In fact, shortages are
widespread in oil field tubular goods. This shortage and the shortage
of drilling rigs are now the limiting factors in oil and gas exploratory
activity.[14] The increased drilling activity has not yet been reflected
in new proven reserves. Industry experts state that it takes three to
seven years to develop new reserves, and it will take some time
before proven reserves increase. Some of the acknowledged experts
state quite positively that domestic exploration and production of oil
and gas will never again be adequate to meet the long-range needs of
this country.[15] The chairman of the FPC, John N. Nassikas, estimates
that "over the next 20 years, there will be an inadequate domestic
supply of both oil and natural gas to meet the forecasted demand in
the United States."[16]

    Solving the gas supply problem by shifting to alternate sources
is hindered by the time required for research and development and
by shortages of raw materials, such as the water needed to convert
coal to gas. The testimony of Steever, reported earlier, estimated
that not much can be done to ease the domestic production shortages
before 1990.[17]

    Data were presented to the National Commission on Materials
Policy Seminar held at the Department of Commerce on December 1,
1972, which showed comparisons of prices and the magnitude of the
demand-supply gap in natural gas. At that time, the highest imaginable
price of oil per parrel was $12.66. In January 1974, Arab oil was
sold at auction for $25 per barrel. The price of oil is a function of

FIGURE 1

United States Gas Supply Versus Estimated Gas Demand, 1970–85

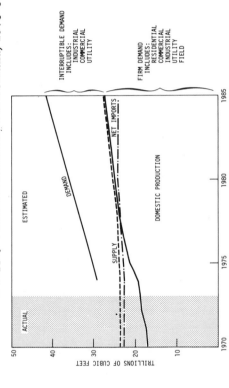

Sources: Future Requirements Committee, Future Gas Consumption of the United States, a report, vol. 5 (Denver: Future Requirements Agency, University of Denver Research Institute, November 1973), pp. 1–8, 50–56; U.S. Federal Power Commission, National Gas Supply and Demand, 1971–90, staff report no. 2 (Washington, D.C.: Bureau of Natural Gas, February, 1972), p. 3; Edwin F. Hardy, manager, Gas Supply Department, American Gas Association, "Supplemental Gas Supply Projections," September 19, 1974; American Gas Association, Gas Supply Committee, Gas Supply Review, supplement, May 15, 1974, p. 2; and interview with Edwin F. Hardy, manager, Gas Supply Department, American Gas Association, September 20, 1974.

37

TABLE 3.3

Estimated Cost of Gas Supplied to the New York Area from New Supply Sources
(in dollars or millions of U.S. dollars)

Supply Source	Gas Cost ($/MM BTU)	Facility Cost ($ mil./1,000 MM CF/D)	Equivalent Energy Cost (oil/$ per barrel)
Arctic gas	1.56 1.56	1,490	9.84
SNG from naphtha (from Canada)	1.60	1,500	10.08
SNG from coal	1.43	1,960	9.01
Methanol (from Persian Gulf)	1.50	1,350[a]	9.45
LNG (from Persian Gulf)	2.01	550[a]	12.66
LNG (North Star)	1.40	—[b]	8.74

[a]Excludes the cost of ships.
[b]Barter trade.

Sources: Compiled from data presented during a December 1972 conference on energy held at the U.S. Department of State; U.S. Federal Power Commission, "Future Gas Supplies from Alternate Sources," Preliminary draft of Chap. 10, National Gas Survey, 1974; and Ted Wett, "SNG from Coal Involves Big Projects," Oil and Gas Journal (June 25, 1973): 131-34.

politics at this time, and industry experts refuse to speculate on what the stabilized price will be. All agree, however, that the current price is unrealistic and will force consumption into other sources of energy if it is maintained for any great period of time.[18] Figure 1 provides an estimate of the gas demand of the various categories of customers and shows that projected production through 1985 is unlikely to satisfy demand. Gas producers probably will not have supplies to service industrial, commercial, and utility consumers currently being served on an interruptible demand basis. The costs of generating gas via the various alternatives are depicted in Table 3.3. These cost data are compared with oil equivalents in terms of price.

The demand estimates confirm discussions with industry representatives and agree with statements by government policy makers that whatever supplies of gas are generated, they will be insufficient to meet demand.[19] Also, the price of North Star gas is cost competitive with alternative sources for gas production.

SUMMARY

In this chapter, an analysis of the U.S. natural gas supply problem was made, and it was determined that a shortage of natural gas exists and is forecast to continue for 20 years. New major sources of natural gas would require three to seven years before production could be brought on-line. The cost of generating gas from alternative sources, coal and petroleum distillates, was expensive, and difficulties were compounded by lack of technology and a shortage of raw materials.

The U.S. natural gas shortage is real and is expected to persist through the 1980s; however, the shortage is due to pricing policies and to unavailability of technology for economic utilization of alternative energy sources. Deregulation of gas in the U.S. domestic market or insistence on a high price for LNG would expedite discovery of alternative resources for natural gas. This study assumes that gas prices will not be held artificially high and that technological breakthroughs will not occur before the decade of the 1990s.

NOTES

1. George Dewan, "Professors Vindicated by Energy Crisis," Washington Post, January 1, 1974, p. D-9.

2. American Gas Association, Natural Gas Supply Problem: Background Report (Arlington, Va.: American Gas Association, 1972), p. 3.

3. Ibid.

4. U.S. Federal Power Commission, National Gas Supply and Demand, 1971-90, Staff Report No. 2 (Washington, D.C.: Bureau of Natural Gas, February 1972), pp. 9, 136.

5. Interview with Charles Krautler, public relations officer of the Washington Gas Light Company, January 7, 1974.

6. American Gas Association, op. cit., pp. 3-6.

7. Krautler, interview.

8. U.S. Federal Power Commission, Natural Gas Reserves Study, National Gas Survey Staff Report, revised (Washington, D.C.: Federal Power Commission, September 1973), p. 1.

9. Ibid., p. 3.

10. Ibid.

11. U.S. Congress, House, Committee on Science and Astronautics, Subcommittee on Energy, Energy Research and Development—An Overview of Our National Effort, Hearings before the Subcommittee on Energy, 93rd Cong., 1st sess., 1973, pp. 1-22.

12. George M. Bennsky, "World Trade in LNG: An American Viewpoint," remarks before the Fourth International Conference on Liquefied Natural Gas at Palace of Nations, Algeria, June 24, 1974, p. 4.

13. John N. Nassikas, "The Role of Liquefied Natural Gas in U.S. Energy Policy," remarks before the Fourth International Conference on Liquefied Natural Gas at Palace of Nations, Algeria, June 24, 1974, p. 5.

14. S. David Freeman, "The Forgotten Energy Agencies," Washington Post, February 10, 1974, p. B1; and Ray, interviews and correspondence.

15. Jack H. Ray, president of the Tennessee Gas Transmission Company, interviews and correspondence.

16. Nassikas, loc. cit., p. 9.

17. U.S. Congress, House, loc. cit., pp. 1-22.

18. Stephen M. Aug, "Arab Prices a Mystery to Exxon Chief," Washington Star-News, January 12, 1974, p. C-back page.

19. Nassikas, loc. cit.

4

THE NORTH STAR
JOINT VENTURE PROJECT

The oil and gas industry is closer to the energy problem than the American public because it has been confronted with dwindling supplies and faced with an obligation to provide a reliable supply of energy to its customers. In view of the great uncertainty of the various supply alternatives, most corporations have hedged and tried to develop several alternative sources simultaneously. The North Star project was generated by the efforts of Tenneco Inc. as one of the actions to be taken to help solve the natural gas supply problem.

TENNECO INC. STRATEGY FOR INCREASING
NATURAL GAS SUPPLIES

The Tenneco strategy to increase supplies of natural gas includes simultaneous actions to accelerate worldwide exploratory operations, foreign gas purchases, and manufacture of synthetic natural gas (SNG) from coal and petroleum distillates. Details of these operations are described below.

Coal and Petroleum Gasification

The Tennessee Gas Pipeline Company has joined five other companies in funding the development of a pilot plant for the conversion of coal into pipeline quality gas and synthetic crude oil. One billion tons of coal have been purchased to manufacture synthetic gas, and Tenneco has optioned a group of North Dakota-Montana coal leases and is in the process of determining the extent of reserves.[1]

Also, Tenneco is planning to construct three plants to manufacture SNG from petroleum liquids and is evaluating several possibilities for the importing of methanol.[2]

Worldwide Exploratory Effort

Tenneco has accelerated wildcat and exploratory drilling on a worldwide scale in the hope of developing new energy sources, with considerable success. In Ethiopia, a test well revealed a high volume of natural gas in the Ogaden Desert, and two wells at another location found signs of oil and gas. In Peru, the first of two wells, drilled about nine miles off the coast, yielded a combined test flow of over 5,000 barrels of oil a day and substantial quantities of gas distillate. In the Red Sea, Tenneco has participated in the drilling of five wells, three of which found signs of oil and gas. In the North Sea, six wells have been drilled and oil discovered. In Spain, a four-well drilling operation is under way about 100 miles southeast of Madrid. Also, Tenneco has purchased exploration rights involving 9 million acres in Thailand and 19 million acres in Indonesia.[3]

Canadian Arctic Islands Gas

Tenneco Oil and Minerals, Limited, was organized in 1955 to develop a supply source of oil and gas in Canada. Recently, this company, together with three other U.S. corporations, formed the Special Arctic Group to conduct wildcat and development drilling in the area of 75 million acres of the barren arctic islands of Canada's Northwest Territories. A $75 million, five-year exploration program is under way. A series of major discoveries has already established sizeable reserves of gas. When sufficient quantities of natural gas are found, a large-diameter pipeline will be built to export the gas from the Arctic to eastern Canada and, hopefully, to the United States. Proposed pipeline routes have been surveyed, and the estimated cost is $8 billion (1974 dollars) for the most difficult construction project ever attempted. This pipeline project, from the standpoint of cost and risk, is beyond the capability of any one company; therefore, a consortium has been formed to develop the research and technology necessary to bring the project to fruition.[4]

ν9২-৪৮৫

TABLE 4.1

Foreign Sources of Natural Gas
(in trillion cubic feet)

Country	Proven Reserves	Ultimate Reserves
Algeria	105	105
Nigeria	40	90
Russia	550	2,100

Sources: North Star Project, a project of Tenneco Inc., Texas Eastern Transmission Company, and Brown and Root (Houston, 1972); and Jack H. Ray, "The North Star Soviet Liquefied Natural Gas (LNG) Import Project," in U.S. Congress, House, Committee on Banking and Currency, International Economic Policy, Hearings before Subcommittee on International Trade on H.R. 774, H.R. 13838, H.R. 12839, H.R. 13840, 93rd Cong., 2nd sess., 1974, pp. 220-63. *H 241-15.7*

395 Wallington . 2 floor.

Liquefied Natural Gas (LNG)

In seeking overseas sources of gas supply, Tenneco's goals were to locate a source within 6,000 miles of the East Coast of the United States capable of supplying 2 billion cubic feet of gas daily (BCFD) for a 25-year period.[5] Three foreign countries had reserves that would satisfy this requirement, as shown in Table 4.1.

Considered in the economic analyses were costs of transportation and estimated availability of gas based on estimated domestic needs of the exporting country. For example, Iran has gas reserves in excess of both Algeria and Nigeria. The cost of transportation would have increased delivered price at Philadelphia; it also tended to further concentrate U.S. energy dependence on the Middle East. In addition, importing energy on the level of the U.S. domestic demand forecast for the decade of the 1980s would have a detrimental effect on the U.S. balance of payments. Therefore, a gas import project structured on the basis of 100 percent barter trade would be highly desirable. Of the three sources listed in Table 4.1, the USSR appeared best suited to satisfy the U.S. needs. In order to be cost competitive in the U.S. domestic market, however, the Soviet gas would have to be imported at a cost no greater than $1.25 per million British thermal units (BTU) — 1,000 cubic feet. It was estimated that gas could be delivered at that cost, based on $0.32 field gathering and pipeline, $0.28 LNG plant operation, and $0.65 LNG tanker incremental costs.[6]

THE NORTH STAR PROPOSAL

The North Star project was initiated by U.S. corporations and was proposed to the Soviets about six months before the United States commenced trade discussions with them.[7] The proposal specified:

1. No dollars could leave the United States.
2. All U.S. dollar revenues acquired from the sale of gas must be used to purchase U.S. goods and services.[8]

The Soviets experienced an annual hard-currency trade deficit of approximately $250 million during the decade of the 1960s. They were anxious to continue trade with the West and particularly to acquire U.S. technology to improve the consumer sector of their economy. They had few items to trade; however, they had an abundance of natural resources. Unfortunately, the bulk of the resources was located in the Siberian or Central Asian areas and was unavailable because the Soviets lacked the finances and technology to extract them within the desired time. The proposed North Star project promised earnings of approximately $10 billion, after debt service over the life of the project, to support the desired trade with the West. When it became obvious that the United States was serious about trade prospects, the Soviets responded favorably to North Star, but specified:

1. No claims could be made on the already overextended Soviet economy, as reflected in the current Five-Year Plan.
2. The project must be completely financed and equipment provided by the West, from gas wellhead to export delivery point.
3. The Soviets desired to own and operate one-half of the LNG tankers required to export the gas.[9]

Although there remained many difficulties to be resolved, the project appeared feasible and promised the following advantages to the United States:

1. U.S. proved gas reserves would be augmented by 18.3 trillion cubic feet (TCF) and would provide 2 BCFD to the East Coast over a 25-year period.

2. There would be no adverse effect on the U.S. balance of payments from the import of the gas.
3. The base of U.S. energy imports would be broadened, providing favorable national security advantages.
4. Soviet earnings from North Star would provide the means for expanded U.S.-Soviet trade.
5. Purchase of equipment for the development of North Star would create 250,000 man-years of jobs in the United States.
6. Increased commercial contact should improve political relations between the two countries.[10]

COMPONENTS OF THE NORTH STAR PROJECT

The North Star project proposes to ship natural gas from North Central Siberia to the Philadelphia area at a rate of 2.1 BCFD for 25 years. The project would be a joint undertaking by three U.S. corporations: Tenneco, Texas Eastern Transmission, and Brown and Root.

The gas source would be the Urengoi gas field in the Tyumen Arctic region, and gas would flow via large-diameter pipeline to a liquefaction plant located on the Kola peninsula at Petsamo, about 100 miles west of Murmansk. There, it would be chilled to - 260 degrees Fahrenheit, at which point it would condense into a liquid form, occupying one six-hundred-fiftieth (0.00153) of its gaseous volume. The LNG would be loaded aboard cryogenic tankers and shipped to a terminal site in New Jersey on the Delaware River across from Philadelphia, Pennsylvania. On arrival, the liquid would be regasified and entered into the U.S. pipeline system to supply East Coast customers.[11] Map 1 graphically portrays the proposed route. The main components of the North Star project are discussed below.

The Urengoi Gas Field

The basic question which comes to mind is: Does this field have the capacity to supply gas at a rate of 2 BCFD for 25 years? The answer appears to be an unqualified yes. Soviet, United States, and international sources confirm that Urengoi is the largest gas field in the world; in fact, with a minimum of 141 TCF of proven reserves (some recent developments indicate the field has 210 TCF), it is

MAP 1

The North Star Project

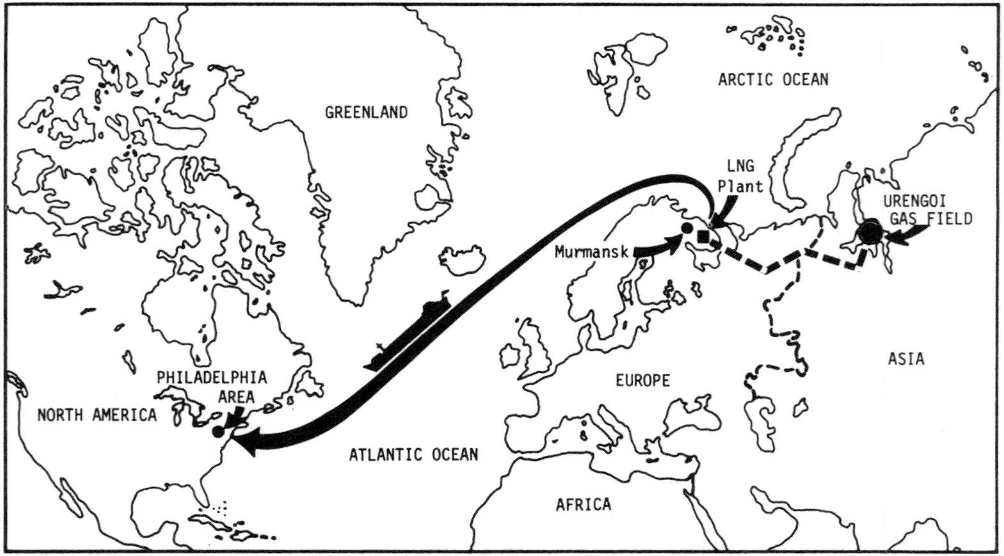

Source: North Star Project, a project of Tenneco Inc., Texas Eastern Transmission Company, and Brown and Root (Houston, 1972), p. 4.

more than twice as large as the largest gas field outside the Soviet Union, the Gronigen gas field in the Netherlands and West Germany (65.3 TCF).[12] The Urengoi field contains quantities of gas equivalent to between 50 and 80 percent of total U.S. proved reserves.[13]

The American Association of Petroleum Geologists scheduled a symposium in April 1968, at which international participants presented papers describing the giant gas and oil fields of the world. A total of 79 giant gas fields was identified, each of which had estimated recoverable reserves in excess of 3.5 TCF. Urengoi was the largest field, with estimated recoverable gas reserves of 210 TCF, plus 1 to 5 million barrels of natural gas liquids.[14] A Soviet publication of this period listed the economically recoverable reserves as 60 TCF and ultimate reserves as 92 TCF.[15] Most recent Soviet sources have increased the commercially recoverable reserves to 141 TCF.[16] The gas field contains additional reserves, but they are not proven by drilling or are not economically recoverable under today's technology. Soviet sources state that the twenty-second, twenty-third, and twenty-fourth Communist party congresses placed enormous emphasis on the role of gas in the energy plan of the Soviet economy.[17] To achieve the energy goals, an extensive exploration was conducted of the Urengoi field. Detailed data were compiled on gas and petroleum distillates from a "deep drilling" experiment to 2,800 meters. Flow, temperature, and pressures were measured, and gas and distillate submitted to laboratory analysis to determine quality standards.[18] Tenneco conducted its own survey and reported the field as shown in Table 4.2.

In addition to the formal study, Tenneco personnel have visited the Western Siberia oil and gas fields, and it is planned to put pipeline experts on the ground for two surveys: one in the forthcoming winter and one in summer. Jack H. Ray, president and project officer for North Star, stated during an interview that he had visited Russia 20 times, with the last visit occurring in June 1974. Some 30 to 40 wells have been drilled at Urengoi, and Ray is convinced beyond any doubt that adequate reserves exist to support North Star.[19]

The evidence is convincing that the Urengoi gas field contains at least 141 TCF of recoverable reserves. Since North Star would require only 28 TCF to satisfy its 25-year delivery requirements, the Soviet government would commit approximately 20 percent of the proven field capacity to export if it did, in fact, go forward with the North Star contract.

TABLE 4.2

Characteristics of the Urengoi Gas Field

Length	108 miles
Width	31 miles
Area	1.06 + million acres
Proven recoverable natural gas	141 + trillion cubic feet[a]
Average production, each well	28 million cubic feet daily
Estimated cost of each well	$1.0 million
Average formation thickness	114 feet
Average porosity	27 percent
Average permeability	550 millidarcys
Average field production rate[b]	2.9 billion cubic feet daily

[a]Recoverable gas reserves are 61.2 percent of the gas in place in the reservoir if the economic limit is set at 28 million cubic feet per day per well at a flowing wellhead pressure of 500 pounds per square inch gauge.

[b]Average rate required for North Star. Urengoi field is estimated to be capable of producing at eight to ten times this rate.

Sources: North Star Project Geological and Reserves Report (Houston: Tenneco Oil Company, March 31, 1972), pp. 1-45; and North Star Project, a project of Tenneco Inc., Texas Eastern Transmission Company, and Brown and Root (Houston, 1972), p. 5.

The Pipeline

North Star would require a pipeline gathering system at the Urengoi field to gather the gas, remove the entrained moisture therefrom, and deliver it to the main transmission line. A main transmission line (48-inch pipe) is required to transport the gas from the field to the liquefaction plant at Petsamo, a distance of 2,600 kilometers (approximately 1,600 miles). Map 1 shows the role of the pipeline in the proposed route. The pipeline route is located mostly in permafrost and would have the following characteristics:

1. The route requires four major river crossings and a 44-mile crossing of the White Sea.

2. The line includes nine compressor stations initially. Each station would have two 32,550 horsepower gas turbine driven compressors. Nine additional stations would be required to fully power the line (optimize throughput).

3. The pipeline system is designed to deliver 2.54 BCFD to the liquefaction plant. The additional nine stations could bring the line volume up to 3.34 BCFD.

4. All equipment to be sold is also available for purchase by the Soviets from other foreign nations, principally France, West Germany, or Japan. The cost to the Soviets in 1980 prices is estimated to be $2.2 billion.[20]

Liquefaction Facilities

Facilities would include a plant, storage tanks, and docking and loading facilities able to load 2.2 BCFD, with the following characteristics:

1. The plant would contain nine liquefaction units, each able to liquify 260 million cubic feet (MCF) of gas per day. (Impurities would be removed from the gas at the plant.)

2. Storage capacity would be 3.6 billion barrels, to be provided by six double-walled steel tanks (206 feet in diameter by 150 feet high), each with a storage capacity of 600,000 barrels.

3. Docking and loading facilities would be able to handle three LNG tankers at berth and to load two ships simultaneously. Each ship could be completely loaded via gravity feed from the storage tanks in a period of 15 hours.

4. All equipment to be purchased by the Soviets is also available from sources outside the United States. The cost of the liquefaction facilities, for U.S. goods and services but excluding Soviet labor, would be $1.5 billion. The price includes inflation and interest to 1980.[21]

LNG Tankers

North Star would require 20 LNG tankers in order to deliver 2.1 BCFD from Petsamo to Philadelphia (4,000 miles). The ships would have the following characteristics:

TABLE 4.3

Cost Estimate of the North Star Project
(in billions of U.S. dollars)

Item	Cost
Pipelines (USSR)	2.2
LNG facilities (USSR)	1.5
LNG ships	2.6
Total cost	6.3
Receiving terminal (U.S.)	0.4
Grand total	6.7

Sources: North Star Feasibility Study, by Brown and Root, Inc., Tenneco Inc., and Texas Eastern Transmission Corporation, vol. 1: Text (Houston, March 31, 1972), pp. 1-23 (Pipelines); pp. 1-11 (LNG Plant); and pp. 1-14 (Ships); and interviews and correspondence with Jack H. Ray and Dan Walsh, September 1973 to July 1974.

1. The capacity of each ship, 125,000 cubic meters, would be limited by the loaded draft (36 feet) in most shipping approach channels of the East Coast of the United States.

2. The average open sea speed of the LNG tankers was estimated to be 18.5 knots with an additional two days allowed on each trip for bad weather. On this basis, each ship would make a round trip in 22 days and would make a total of 15.6 round trips per year.

3. The cost of each LNG tanker delivered into service between 1978 and 1981 is estimated at $131 million.

4. Six U.S. shipyards are capable of building LNG tankers, and construction time is estimated at 36 months for the first ship and 4 to 6 months for each additional ship. A large number of foreign yards are capable of construction, but they are fairly well booked.[22]

The total project cost is estimated at $6.3 billion, with an additional $0.4 billion for a receiving terminal in the United States, as shown in Table 4.3.

SUMMARY

The North Star proposal was originated by Tenneco Inc. as one of several actions taken in a strategy designed to help solve the natural gas supply problem. The strategy involves simultaneous actions taken to accelerate worldwide exploratory operations, purchase of natural gas from foreign and domestic producers, and manufacture of SNG from coal and petroleum distillates.

The cost of natural gas under the North Star proposal was estimated to be $1.25 per million BTUs, which would make it cost competitive in the U.S. domestic market in the decade of the 1980s. Analysis of the gas reserves of the Urengoi gas field in North Central Siberia showed them to be more than adequate to support the requirement for delivery to the United States of 2.1 BCFD for a 25-year period. The pipeline, liquefaction facilities, and LNG tanker requirements were described, and the total cost was estimated to be $6.7 billion.

A series of advantages to the United States would ensue from North Star. U.S. gas supply would be augmented by the import of 18.3 TCF of gas. Impact on the U.S. balance of payments would be eased because energy would be obtained through barter trade. Soviet dollar earnings would be tied to the purchase of U.S. goods, thus promoting U.S.-Soviet trade and creating, potentially, 250,000 man-years of jobs for the U.S. economy.

NOTES

1. Tenneco, 1972 Annual Report (Houston: Tenneco Inc., 1972), p. 8.

2. North Star Project, a project of Tenneco Inc., Texas Eastern Transmission Company, and Brown and Root (Houston, 1972), p. 1.

3. Tenneco, Special International Issue, vol. 7, no. 2 (Houston: Tenneco Inc., summer 1973).

4. Ibid.; and Jack H. Ray and Daniel S. Walsh, interviews during the period January to July 1974.

5. North Star Project, loc. cit., p. 1.

6. North Star Project Feasibility Study, by Brown and Root, Inc., Tenneco Inc., and Texas Eastern Transmission Corporation, vol. 1, pp. 1-7 (Economics).

7. Interviews with Jack H. Ray and Gennadiy V. Dmitriev, September 1973 through October 1974.

8. North Star Project, loc. cit., p. 2.

9. North Star Project Feasibility Study, loc. cit., vol. 1, pp. 1-5.

10. North Star Project, loc. cit., pp. 2-11.

11. Ibid.

12. Michael T. Halbouty, Geology of Giant Petroleum Fields (Tulsa: American Association of Petroleum Geologists, November 1970), p. 509.

13. U.S. Federal Power Commission, National Gas Supply and Demand, 1971-90, Staff Report No. 2 (Washington, D.C.: Bureau of Natural Gas, February 1972), p. 136.

14. Halbouty, op. cit., p. 509 and Table 2 foldout.

15. Mikhail S. L'vov, Resursy Prirodnovo Gaza SSSR [Natural Gas Resources of the USSR] (Moscow: Nedra, 1969), p. 163.

16. A. P. Agishev, V. G. Vasiliev, and Y. M. Vasiliev, "The Principal Gas-Bearing Areas of the Soviet Union," paper presented at the 12th World Gas Conference, Paris, April 1973, p. 4.

17. Dmitri V. Velorusov, et al., Osvoyeniye Neftyanykh Mestorozhdenij Zapadnoj Sibiri [Mastering the Petroleum Fields in Western Siberia] (Moscow: Nedra, 1972), p. 3.

18. Ibid., pp. 124-38.

19. North Star Project Geological and Reserves Report (Houston: Tenneco Oil Company, March 31, 1972), pp. 1-45; North Star Project, loc. cit., p. 5; and Ray, interviews, September 1973 through July 1974.

20. North Star Project Feasibility Study, loc. cit., vol. 1, pp. 1-23 (Pipelines); North Star Project, loc. cit., pp. 5-6; and Jack H. Ray and Daniel S. Walsh, interviews, September 1973 to July 1974.

21. North Star Project Feasibility Study, loc. cit., vol. 1, pp. 1-11 (LNG Plant); and North Star Project, loc. cit., p. 6.

22. North Star Project Feasibility Study, loc. cit., vol. 1, pp. 1-14 (Ships); and North Star Project, loc. cit., pp. 6-7.

II

INSTITUTIONAL OBSTACLES TO THE NORTH STAR PROJECT WITHIN THE SOVIET SYSTEM

The Soviet policy of scrupulously meeting terms of commercial contracts and promptly paying debts since World War II has earned for the USSR a reputation as the "most reliable debtor in the world."[1] Nevertheless, this preferred risk rating with international and U.S. bankers is under review. The current international monetary uncertainty is causing some uneasiness, but a more deep-seated reason is the concern about the long-range impact of debt service on the Soviet balance of payments.[2]

Russia has been paying for the importation of Western technology largely with the shipment of oil. In fact, in 1973 the sale of Soviet oil to the West resulted in a $990 million increase in earnings.[3] Soviet exports of crude oil to the West, however, are expected to peak in 1976 and then decline as the result of domestic constraints on supply, higher domestic consumption, and commitments to Eastern Europe.[4] Furthermore, petroleum prices may be reduced in the future. An increase in gasoline inventory stocks in the United States was recorded in May 1974 for the first time in 20 years, and the upper-range prices for gasoline and distillate fuel oils had decreased due to abnormally high inventory accumulations.[5]

With the expected decrease in the export of oil, the Soviet Union has the option of paying for the desired technology with natural gas, which is found in abundance in several areas of the Soviet Union. In fact, the largest potential reserves of natural gas in the world are located in the Urengoi gas field located in Western Siberia.

In Part II, the North Star proposal for the delivery of gas to the United States is analyzed in the context of institutional constraints, resource limitations, state of technology, and economic and financial resources of the USSR. The implications of Russian ideology are described to provide an understanding of the area of compromise which is feasible in working out joint business arrangements. Soviet planning objectives and the bureaucratic structure used to implement policy related to the Soviet energy plan are presented, followed by an analysis of the Soviet gas supply-and-demand projections, to determine whether the quantities of gas proposed for export under North Star are feasible. Next, the problem of technology transfer is analyzed to determine whether debt service could take on the added burden of North Star and, conversely, whether pressure on the balance of payments generated by North Star would encourage the Soviets to maintain deliveries under the project.

NOTES

1. Business International, S.A., Doing Business with the USSR (Geneva, November 1971), p. 92.

2. Ibid., p. 95, and a series of private interviews with Chase Manhattan Bank and First National City Bank of New York during September 1973. Bank officials indicated that the "holiday was over" and that Soviet requests for loans would be handled as preferred customers but that reduced interest rates and special preference treatment would not be granted.

3. Christopher S. Wren, "Russian Oil Profits from the West Climb," New York Times, June 5, 1974, pp. 1, 65.

4. U.S. Congress, Joint Economic Committee, Subcommittee on Priorities and Economy in Government, Allocation of Resources in the Soviet Union and China, Hearings, 93rd Cong., 2nd sess., April 12, 1974, p. 24.

5. Gregory J. Shuttlesworth, "Highlights for May 1974," The Chase Manhattan Bank monthly review, The Petroleum Situation (June 24, 1974): 1.

5

SOVIET ORGANIZATIONAL
STRUCTURE
AND IDEOLOGY

The North Star project is designed to operate in two totally different environments: the planned economy and the free-market system. Institutional structures and operating procedures in each of the systems differ, and the cold war structural barriers, which curtailed commercial relations between the Soviet Union and the United States, precluded the development of an infrastructure which might promote expanded trade. This chapter deals with institutional structures and procedures employed in the Soviet planned economy.

INSTITUTIONAL STRUCTURES OF THE
USSR PLANNED ECONOMY

The planned economy, by definition, does not produce a surplus. Every item of production and every item of consumption are planned. Theoretically, the last grain of wheat and the last gram of steel are to be consumed by the end of the year. The Five-Year Plan and annual adjustments direct how production is managed and when consumption occurs. Prices and costs do not determine, by themselves, the economic decisions of the Soviet government. Cost, profit, utility, indifference curves, and the entire stable of economic concepts, which are so essential for successful performance in a free-market economy, are either not applicable or may have a different application and meaning in a planned economy. Projects with a very high return on investment are often ignored by the Soviets, while other projects that promise only losses in a free-market sense are enthusiastically pursued. The Westerner who thinks only in free-market concepts when working with a Soviet counterpart is foredoomed to failure. In this sense, he is hampered by the very techniques that have made him a success in his own environment.

The North Star proposal, in its basic essence, is different from existing commercial arrangements. It is not an investment, because equity ownership is not conveyed. It is not a long-term lease or sale because equity rights are not involved. It is not conventional barter trade, because it spans a 25-year period, which is far beyond conventional barter in today's commercial environment. The North Star proposal, however, does have characteristics that resemble parts of all the above commercial arrangements.

In recent years, Soviet foreign trade has enjoyed a higher priority than domestic consumption, particularly with regard to the Soviet allocation of energy resources. Within the U.S. Department of the Interior, there was much speculation and debate in 1968 about whether Soviet gas shipments would be made to Austria that year because of the critical shortages known to have existed in the domestic economy. Deliveries made to Austria demonstrate that terms of contracts were met at the expense of the domestic economy.[1] (Also see Table 6.6.)

In the free-market system of the United States, cost and price determine production and consumption. Surplus is a normal result of supply-and-demand adjustments. Price determines consumption. Goods are often shipped to foreign consumers in spite of existing domestic demand as a response to price pressures. U.S. export restrictions on lumber and soybeans during 1973 reflected government intervention when price pressures became extreme.

The North Star project, if agreement is reached, will engage the Western businessman into the minutiae of the Soviet planned economy. Although the equipment and much of the technology for North Star will come from the United States, the installation of the equipment on site will rely on Soviet labor, and the transportation of equipment to various sites within the Soviet Union will rely on Soviet logistic support.[2]

SOVIET IDEOLOGY

Every man, whether he is a government official, businessman, or artist, is influenced by his cultural heritage. Behavioral traits, such as the American obsession with time, the Latin's almost total insensitivity to time, or the Soviet penchant to unquestioned obedience to orders from above, are familiar characteristics to even the occasional international traveler. An effective manager takes these factors into account when planning social or business arrangements. The subtle ideological factors, which influence the intellectual thought and actions of foreigners, are much less obvious, but they are, perhaps, more important. This aspect of management is often

overlooked in U.S. government relations with the Soviet Union, and
more so in business relations.

Ideology implies rigidity, tightly held views, with no area of
compromise. The series of agreements signed in 1972[3] demonstrate
that the United States and the Soviet Union, recent archenemies in
cold war, have moved toward detente. This movement began during
the Vietnam War and would have been considered treasonous a few
years earlier. Yet, each nation seems willing to compromise to
achieve desired goals. Their mutual actions reflect a hope of finding
solutions to very serious problems which plague their respective
economies.

For the United States, persistent inflation, unemployment,
balance-of-payments deficits, and shortages (particularly in energy)
threaten serious recession, or even prolonged depression.[4] Cooperation with the Soviet Union suggests possible solution, or at least a
mitigation of the more serious aspects of the problems.[5]

For the Soviet Union, persistent agricultural shortages, a lagging
growth rate, unsatisfactory productivity, backward technology, and a
shortage of investment funds have caused and continue to cause a
failure to achieve established goals[6]—a failure which can lead to the
downfall of the leaders. Cooperation with the United States offers
possible solutions, in some cases the only feasible solution within
the time constraint desired, to the problems that plague the Soviet
economy.[7]

Extensive Soviet energy and raw material endowments, as con-
trasted with U.S. agricultural abundance and technological superiority,
suggest that mutual benefits could result from U.S.-Soviet cooperation.
Customs, tradition, and legal and institutional procedures within the
two economic systems differ. There is much discussion in both
countries concerning whether the two systems are converging; that
is, are they becoming more alike, or are they so fundamentally different that they can never be reconciled? The implication is that
convergence would facilitate mutual cooperation.

The concept of convergence gradually evolved in the post-Stalin
era, and no single person can be associated with the theory. The
concept was described in an article published in 1961.[8] The general
thesis holds that large, complex, technocratic economies must adopt
the same managerial procedures in order to permit the economy to
function efficiently. Thus, since they converge toward the same goal,
they tend to become like each other. A major academic dispute
erupted with the publication of Political Power: U.S.A./USSR.[9] The
authors, Zbigniew Brzezenski, of Columbia University, and Samuel
Huntington, of Harvard, held that the United States and the USSR
were not becoming alike. In a critical book review, Gabriel Almond,

of Stanford University, took issue with the published findings, but was forced to retract his criticism after a sharp defense by the authors.[10] Soviet writers joined in the argument, strongly disagreeing with the concept of convergence, stating that it had very little validity, ". . . barely enough to put in a thimble."[11]

It is the view of this study that both Soviet and U.S. systems are changing to conform to the requirements of modern technology. In this sense, they are changing on a parallel course. They adapt to technology only as far as is necessary, but each refuses to give up individual aspects of its cultural heritage. However, because both are changing to conform to the same technological requirements, mutual cooperation is facilitated. Whether the paths do or do not converge in the distant future is immaterial.

The Soviets have established a legal basis for cooperation with the West by formal approval of the instructions for implementing the 1974 plan for development of the Soviet economy:

> The plan provides for the further development of economic ties between the USSR and the developed capitalist countries. Trade arrangements with these nations will make it possible for the national economy to obtain modern machinery and equipment, as well as other goods not manufactured in the USSR. Besides this, the Soviet Union will continue the practice of soliciting foreign investments for the purpose of developing various industrial enterprises in the USSR, and to repay these investments by using part of the output of these enterprises.[12]

If projects such as the North Star are to succeed, cooperative arrangements must take into account the rigidities of each system. The foremost criterion for success on a long-term basis is to assure mutual benefits to both parties. The project itself must rest on a sound economic foundation. The question can be asked whether, on this basis, ideological rigidities will permit the kind of compromises that must be met to implement the North Star project. The answer to this question resides in the philosophical underpinnings of the two societies.

Both the planned economy of the Soviet Union and the market economy of the United States operate on the basis of organizational institutions and operating procedures which are devised to achieve the policy goals of their respective societies. Cultural roots for both structures extend back to Judeo-Christian origins associated with Jewish prophet kings and continuing through Greek democracies and the Roman empire. On the Soviet side, the Mongol occupation of

Russia left deep imprints on the national psyche. Present politicoeconomic organizational structures in both countries appear to stem
from the political turmoil of the eighteenth century, which culminated
in the French revolution. This revolution constituted a point of diver-
gence for the two patterns for the organization of society. The prevail-
ing philosophies of that day seem to have influenced the organizational
structures which have evolved for the achievement of politicoeconomic
goals. The Soviet system bears a striking resemblance to the doctrines
of Rousseau, particularly the "General Will":

> Man is born free, and everywhere he is in chains. . . . How
> did this change come about? I do not know. What can make
> it legitimate? That question I think I can answer . . . the
> social order is a sacred right which is the basis of all other
> rights.[13]

Rousseau believed that the major problem of society was to
preserve the social order, that is, to find a form of association
which would protect the person and the goods of each member of
society with the entire force of the community. The solution lay in
the forming of a social pact, whereby the individual surrendered all
his rights to a body politic and placed himself under the general will
of that body. This act of submission created a moral and collective
community with a common identity, unity, life, and will. This general
will became the superior, the guide, and the will of the individual.

Rousseau held that the individual had nothing to fear from the
community because the general will embodied ultimate truth. It
could not be wrong or harmful to the individual because it was truth—
the consensus made it so. However, an individual could be in error,
for he could desire things that were not in his best interest. Therefore, if he did slip into error, the superior will of the community,
the ultimate good, had the obligation to intervene to protect him from
himself. It would force him to be free. In other words, the state
becomes the infallible authority, a divine form on earth. It is better
qualified and more capable of making correct decisions than is the
individual. Therefore, man's real substantive freedom consists of
submitting himself to the higher rationality of state. Only he who
obeys the state is free:

> . . . the Sovereign [community] being formed wholly of
> the individuals who compose it, neither has nor can have
> an interest contrary to theirs . . . merely by virtue of
> what it is, is always what it should be . . . this, however,
> is not the case with relation of the subjects . . . each

individual, as a man, may have a particular will contrary
or dissimilar to the general will which he has as a citizen.
. . . In order that the social compact may not be an empty
formula, it tacitly includes the undertaking, which alone
can give force to the rest, that whoever refused to obey
the general will shall be compelled to do so by the whole
body. This means nothing less than that he will be forced
to be free.[14]

The rationale of forcing one to be free has been extended by the
Soviets to the international arena. The concept reappears clearly in
the justification for the Soviet invasion of Czechoslovakia in 1968:

. . . "self-determination," that is the separation of
Czechoslovakia from the Socialist Commonwealth, would
run counter to Czechoslovakia's fundamental interests
and would harm other socialist countries. Such self-
determination, whereby NATO troops could approach the
Soviet borders, and the Commonwealth of European
socialist countries would be dismembered, in fact, infringes
on the vital interests of these countries and fundamentally
contradicts the right of these peoples to self-determination.
The Soviet Union and the other socialist states, in fulfilling
their internationalist duty to the fraternal peoples of Czechoslovakia, had to act and did act in resolute opposition to the
anti-socialist forces in Czechoslovakia.[15]

This concept, forcing Czechoslovakia to be free to accept the
blessings of socialism, came to be known as the Brezhnev Doctrine.
It is accepted by party ideologues as a fundamental truth and fully
justifiable action. A Westerner tends to view it as a thinly veiled
excuse for intervention.

Rousseau's concept of the general will is embodied in the Soviet
decision-making and policy-implementing procedure called "Demo-
cratic Centralism." Theoretically, this structure permits sugges-
tions and discussions concerning policy issues to be forwarded
upward from party cells, through intermediate organizational units
to the Politburo, where decisions are made. The Politburo is the
nerve center of the system. It is the locus of all knowledge and all
truth. Its decisions are final and are to be carried out without
question or hesitation. Ideally, this mechanism permits policy to be
formulated democratically with all participants free to add their
views, and the mechanism permits policy to be executed centrally
for maximum efficiency. In practice, decisions are often made

centrally, without consultation, and are ordered to be implemented without comment or question. The North Star project is intimately affected by this institutional structure and procedure.

This mechanism for guiding the economic activities of the nation was acted out in December 1973, when the Plenary Session of the Central Committee of the Communist party unanimously approved the draft plan for the development of the Soviet economy during 1974, and the budget to implement it, with these instructions:

> To completely and totally endorse the activity of the Politburo to carry out the domestic and foreign policy decisions. . . .
>
> . . . to recognize the decisive importance of fulfillment of the plan to create the conditions necessary to improve the people's well-being and cultural level. . . .
>
> . . . to direct Union-Republic, Territory, Province, and all Party Committees to mobilize all workers to fulfill the 1974 national economic plan under the guidance and direction of Comrade L. I. Brezhnev.[16]

After approval of the plan by the party organs, it was approved by the Supreme Soviet and carried out by the formal governmental structures. These structures resemble in external form and function similar institutions of the other nations of the world. However, in internal functioning, they lack the decentralized decision-making capability and flexibility of their Western counterparts.

In order to embark on a cooperative arrangement as extensive as North Star, members of both economic systems must be willing to compromise. The period of the cold war reflected an era of extreme ideological rigidity. Both parties wore ideological blinders. Each was frozen in preconceived precepts and refused to budge from its ideological position.

In the decade of the 1960s, cold war positions began to ease. Western Europe expanded commercial relations with Eastern Europe and the Soviet Union, but the United States remained aloof. Soviet trade with the free world, mostly Western Europe, increased from $7.4 to $19.8 billion. However, trade volume between the United States and the USSR remained relatively unchanged during the decade.[17]

In the 1970s, there was a sharp change in U.S. policy. Trade barriers were relaxed, and by 1973 the U.S.-Soviet total trade turn-around reached almost $1.5 billion.[18] This new volume represented an increase of almost 800 percent over the average performance of the 1960s. Most of this trade was conducted under terms of cash, barter, or short-term financing. However, the Export-Import Bank did grant six loans, in the amount of $178 million, and this amount was matched by equivalent loans granted by various U.S. banks. The six projects were financed on the basis of 10 percent Soviet contribution, 45 percent U.S. bank loans, and 45 percent Export-Import Bank loans. The loans will be repaid in varying periods from 7 to 12 years.[19]

The North Star project proposes a larger and deeper commitment, approximately $2 billion covering 15 years, than has heretofore been granted in U.S. export-import financing.

The ideological factors, which will influence the North Star decision, are reflected in Soviet legal and institutional structures. The important legal justification by the Soviet Union has been granted. The Politburo approved and the Supreme Soviet endorsed negotiations for foreign assistance in the development of the Arctic natural resources.[20] The 1974 plan for the national economy, however, has assigned tasks and has allocated resources for the coming year. To begin action on North Star in 1975 would require the almost impossible task of reallocating these resources.

SUMMARY

In this chapter, an examination was made of the Soviet institutional structures and operating procedures which might pose obstacles to implementing the North Star project. The findings are

1. Soviet ideological constraints exist and will continue to be a factor throughout the planning and implementation of North Star. The Soviet Union appears to be willing, however, to bend its ideological constraints to the extent necessary to accommodate the project.

2. The legal foundation, in the sense of a formal document approved by the Communist party and the government, has been enacted into law in the Soviet Union.

NOTES

1. Discussions held in a series of private interviews with Dr. V. V. Strishkov, Soviet specialist, Division of Fossil Fuels, U.S. Department of the Interior, concerning Soviet mineral and fuels industries, between November 1973 and March 1974.

2. North Star Project Feasibility Study, by Brown and Root, Inc., Tenneco Inc., and Texas Eastern Transmission Corporation, vol. 1, p. 2 (Capital).

3. U.S. Department of Commerce, U.S.-Soviet Commercial Agreements, 1972.

4. Allied Social Sciences Convention, New York, December 27-30, 1973. This theme was consistently repeated at approximately twenty conference sessions attended by this writer during the convention.

5. Steven Lazarus, testimony in U.S. Congress, Joint Economic Committee, Soviet Economic Outlook, Hearings, 93rd Cong., 1st sess., July 19, 1973, pp. 112-14.

6. K. M. Grasimov, "O Gosudarstvennom Plane Razvitia Narodnovo Khozyaistva SSSR na 1974 God" [Concerning the Government Plan for Development of the National Economy of the USSR in 1974], Izvestia, December 15, 1973, p. 2.

7. N. K. Baibakov, "O Gosudarstvennom Plane Razvitia Narodnovo Khozyajstva SSSR na 1974 God" [Concerning the Government Plan for Development of the National Economy of the USSR during 1974]. Pravda, December 13, 1973, p. 3.

8. Jan Tinbergen, "Do Communist and Free Economies Show a Converging Pattern?" Soviet Studies, April 1961.

9. Zbigniew Brzesenski and Samuel Huntington, Political Power: U.S.A./USSR (New York: Viking Press, 1963).

10. Gabriel A. Almond, "Book Reviews," American Political Science Review (December 1964): 976, 977; (June 1965): 446, 447.

11. L. Leontiev, "Myth about 'Rapprochement' of the Two Systems," Ekonomicheskaya Gazeta, no. 49 (December 1966).

12. Baibakov, op. cit., p. 3.

13. Jean Jacques Rousseau, "The Social Contract," bk. 1, Chap. 1, in Great Books of the Western World, ed. R. M. Hutchins (Chicago: Encyclopaedia Britannica, Inc., 1952), p. 387.

14. Ibid., bk. 1, Chap. 7, pp. 392-93.

15. S. Kovalyev, "Suverenitet i Internatsionalnye Obyazannosti Sotsialisticheskikh Stran" [Sovereignty and the International Obligations of Socialist States], Pravda, September 26, 1968, p. 4.

16. "Informatsionnoye Soobshcheniye: O Plenume Tsentralnovo Komiteta Kommunisticheskoj Partii Sovetskovo Syuza" [Informational Notes: Concerning the Plenum of the Central Committee of the Communist Party of the Soviet Union], Pravda, December 12, 1972, p. 1.

17. Peter G. Peterson, The United States in the Changing World Economy (Washington, D.C.: Council on International Economic Policy, 1971), p. 27.

18. In an informal interview with Steven Lazarus, Washington, D.C., January 3, 1974, he stated that the United States had become the largest Western trading partner of the Soviet Union in 1973, with exports of $800 million in grain and $500 million in manufactures, and imports of almost $200 million.

19. Export-Import Bank of the United States, press release of March 21, 1973; September 27, 1973; and December 21, 1973.

20. Baibakov, op. cit., p. 3.

6

THE ROLE OF GAS
IN THE SOVIET
ENERGY STRATEGY

The North Star project proposes the shipment of natural gas of 2 billion cubic feet per day (BCFD) for a 25-year period from Western Siberia to the East Coast of the United States. Provisions for this significant quantity of energy must be made in the Soviet energy plan, and detailed arrangements for its export must be included in Soviet economic planning.

SOVIET ENERGY AND NATURAL GAS STRATEGY

A national energy plan was a preoccupation of Lenin almost from the moment the Communists seized power. In the hope of wiping out all elements of the free market to achieve the theoretical advantages of higher productivity of socialism, Lenin nationalized heavy industry, the banking system, foreign and domestic trade, and almost all handicraft and small industry. The disruption caused by these policies brought the economy into a state of chaos. By 1920, a fuel crisis threatened to disrupt the economy. At the urging of Gleb Krzhizanowskij, Lenin proclaimed the first energy plan, the Plan of the State Commission for Electrification of Russia (GOELRO). The plan was to solve Russia's energy problem by general electrification of the economy. It was popularized under the slogan "Communism is the power of the Soviets plus electricity." The plan was formally adopted in December 1921, but no steps were taken to implement it.[1] Partial plans were formulated on a five-year basis, but no real planning was done until 1925. After 1925, annual plans in the form of control figures were drawn up. There were no links between the control figures and the Five-Year Plans, nor were production targets set, but some influence was exerted through control of industrial production.[2]

The first formal Five-Year Plan was implemented on October 1, 1928. Successive plans followed up to the current Ninth Five-Year Plan (except for the 1942-46 period, which was without a plan because of World War II). Energy objectives were carefully outlined in the plans from the very beginning, with the Soviet government publishing statistics annually, as reflected in Table 6.1.

During the period 1928 to the mid-1950s, Soviet energy policy was directed at "mineralization of the fuel balance."[3] Utilization of coal was emphasized, and by 1950 it constituted two-thirds of the energy base of the Soviet Union. During this period, the petroleum contribution to the energy base actually decreased, and natural gas constituted an almost insignificant 2.5 percent of the energy produced. In 1957-58, the first overall evaluation of Soviet gas reserves was made, and reserves were roughly estimated at 19 trillion cubic meters (TCM). At the time, it was estimated that the USSR probably possessed the largest reserves of natural gas in the world.[4] As a result of extensive exploratory and development work in the interim period, the Soviets revised the estimate of gas reserves to 18 TCM in 1973.[5] Russia is thus recognized as having the largest reserves of natural gas in the world.[6]

In mid-1955, Soviet energy strategy shifted. A decision was made to replace coal with gas and oil as the primary fuel. Soviet energy strategy is most easily seen in the statistics published concerning the results achieved under the energy plans. Table 6.2 provides detailed data concerning the energy sources by type in terms of standard fuel quantities and in percentage contribution to the energy base. The shift in energy balance is dramatically shown in the figures for 1955 and 1960. Gas and petroleum rose from 23.5 to 38.4 percent of the energy base, while coal declined from 64.8 to 53.9 percent, as shown in Table 6.2

The Soviet energy plan reflects both the strengths and weaknesses of the planned economy as a mechanism for achieving desired goals. The major strength of the planned system is its ability to concentrate resources and to plan on a long-term basis. Large shifts in the energy base, from firewood to coal and from coal to petroleum and gas, can be achieved relatively quickly. There is, however, an inflexibility in the central plan. Once a decision has been made, it is difficult to change. An inefficient policy tends to remain in effect long after economic factors would justify a change.

Soviet energy policy, until 1950, retained an emphasis on inferior fuels such as peat and lignite. The shift to oil and gas as an energy base in the Soviet Union was late when compared to the experience of Western Europe and the United States.[7] Soviet energy policy during this period can be criticized with hindsight. It must be

TABLE 6.1

The Soviet Energy Plan: Balance of Fuel-Energy Resources
(in standard fuel tons = 7 million kilocalories)

	1913	1940	1960	1965	1970	1971	1972
Resources (total)	64.4	283.6	836.5	1121.5	1399.8	1474.5	1556.5
Production (extracted)							
Fuel	48.2	237.7	692.8	966.6	1221.8	1284.9	1353.8
Hydroelectric energy	0.0	0.6	6.3	10.0	15.3	15.5	15.1
Imports	8.0	3.1	10.7	9.1	14.1	23.4	35.3
Other inputs	2.4	10.2	32.7	35.5	36.5	35.8	35.8
Balance at start of year	5.8	32.0	94.0	100.3	112.1	114.9	116.5
Distributions (total)	64.4	283.6	836.5	1121.5	1399.8	1474.5	1556.5
Consumption	57.6	249.5	678.0	897.8	1117.9	1117.5	1248.8
Of this total:							
Generation electric power	2.0	44.7	212.2	335.0	452.5	476.1	494.4
Technological and other uses	55.6	204.8	456.8	562.8	665.4	701.4	754.4
Exports	1.2	1.1	59.8	116.7	167.0	180.5	187.6
Balance at end of year	5.6	33.0	98.7	107.0	114.9	116.5	120.1

Source: USSR, Central Statistical Directorate, Narodnoye Khozyajstvo SSSR v. 1972 G [The National Economy of the USSR in 1972] (Moscow: Statistika, 1972), p. 70.

TABLE 6.2

Fuel Production by Type

(computed in terms of standard fuel—7,000 kilocalories)

Years	Total	Oil*	Gas	Coal	Peat	Shale	Wood
				Millions of Tons			
1913	48.2	14.7	—	23.1	0.7	—	9.7
1922	29.7	6.7	0.03	9.0	0.9	0.0	13.1
1940	237.7	44.5	4.4	140.5	13.6	0.6	34.1
1945	185.0	27.8	4.2	115.0	9.2	0.4	28.4
1946	202.7	31.0	4.5	127.3	11.2	0.7	28.0
1950	311.2	54.2	7.3	205.7	14.8	1.3	27.9
1955	479.9	101.2	11.4	310.8	20.8	3.3	32.4
1960	692.8	211.4	54.4	373.1	20.4	4.8	28.7
1961	732.7	237.5	70.8	370.1	19.5	5.2	29.6
1962	778.6	266.5	84.6	379.7	12.9	5.8	29.1
1963	847.1	294.7	105.1	388.4	21.7	6.5	30.7
1964	912.2	319.8	127.0	403.3	22.2	7.1	32.8
1965	966.6	346.4	149.8	412.5	17.0	7.4	33.5
1966	1022.1	379.1	170.1	420.1	24.4	7.5	31.9
1967	1088.4	411.9	187.4	428.6	22.4	7.5	30.6
1968	1126.6	442.1	201.2	428.7	18.3	7.6	28.7
1969	1177.4	469.6	215.5	439.6	16.7	8.0	28.0
1970	1221.8	502.5	233.5	432.7	17.7	8.8	26.6
1971	1284.9	537.3	250.6	444.2	16.7	9.5	26.6
1972	1353.8	572.6	264.6	459.8	21.2	9.9	25.7

In Percentage of Total

Year							
1913	100	30.5		48.0	1.4	—	20.1
1922	100	22.5	0.1	30.3	3.0	0.0	44.1
1940	100	18.7	1.9	59.1	5.7	0.3	14.3
1945	100	15.0	2.3	62.2	4.9	0.2	15.4
1946	100	15.3	2.2	62.8	5.5	0.4	13.8
1950	100	17.4	2.3	66.1	4.8	0.4	9.0
1955	100	21.1	2.4	64.8	4.3	0.7	6.7
1960	100	30.5	7.9	53.9	2.9	0.7	4.1
1961	100	32.4	9.7	50.5	2.7	0.7	4.0
1962	100	34.2	10.9	48.8	1.7	0.7	3.7
1963	100	34.8	12.4	45.9	2.5	0.8	3.6
1964	100	35.1	13.9	44.2	2.4	0.8	3.6
1965	100	35.8	15.5	42.7	1.7	0.8	3.5
1966	100	36.7	16.5	40.7	2.3	0.7	3.1
1967	100	37.8	17.2	39.4	2.1	0.7	2.8
1968	100	39.2	17.9	38.0	1.6	0.7	2.6
1969	100	39.9	18.3	37.3	1.4	0.7	2.4
1970	100	41.1	19.1	35.4	1.5	0.7	2.2
1971	100	41.8	19.5	34.6	1.3	0.7	2.1
1972	100	42.3	19.5	34.0	1.6	0.7	1.9

*Including gas condensate.

Source: USSR, Central Statistical Directorate, Narodnoye Khozyajstvo SSSR v. 1972 G [The National Economy of the USSR in 1972] (Moscow: Statistika, 1972), p. 205.

conceded, however, that when the decisions were made, the Russian-proven reserves of gas and oil were low, and the Soviets were not aware of their rich natural resource endowments. In addition, transportation costs influenced policy. The technological breakthrough, which sharply reduced the cost of transporting petroleum and gas by use of pipelines, was not available to the Soviet Union until after World War II. Soviet energy resources were located far distant from consumption areas. Transportation costs were disproportionately high, and rail and transport facilities were overburdened. Finally, the risk involved in oil investment tended to place petroleum at a disadvantage when compared to coal, because the Soviet planners used average cost data in deciding between energy products. Coal had proven reserves and firm development cost figures to aid in planning, while the development costs of petroleum and gas were uncertain and ultimate costs unknown.[8]

Although the Soviet energy plan gave low priority to oil and gas prior to 1950, large investments were made in synthetic development of these fuels. Between 1945 and 1958, approximately 41.7 million rubles were invested in underground gasification of coal and 150 million rubles in the production of gas from shale and coal. Particular emphasis was placed on synthetic liquid fuel plants, underground gasification of coal, and production of artificial gas from shale and coal, but none of these programs was very productive. The Fourth Five-Year Plan (1946-50) proposed the creation of a synthetic liquid fuel industry based on coal and shale. Plants capable of producing 900,000 tons of liquid fuel by 1950 were to be built in Eastern Siberia, Northern Caucasus, and Leningrad Oblast. In the Fifth Five-Year Plan (1950-55), M. Saburov, chairman of the State Planning Commission (GOSPLAN), mentioned plants "in the eastern regions of the country." No mention was made of synthetic liquid fuel in the Sixth Five-Year Plan.[9]

When the Soviet energy policy changed in the 1950s, the Seven-Year Plan (1959-65) shifted emphasis to oil and natural gas as the energy base of the USSR. The inefficient gasification projects had been abandoned, and some of the uneconomic mines were closed.[10] Some of the facilities which had been constructed, however, continued to operate and supplied approximately 1 percent of the gas production in 1970.[11]

The shift to oil and gas in the Soviet energy base has been progressing relatively rapidly. The Ninth Five-Year Plan has set as a goal "to increase the weight of oil and gas in the overall energy balance, by 1975, to not less than 67 percent."[12]

Soviet planners fully understand the potential gains in efficiency inherent in the conversion from coal. These efficiencies result from

the fuel used as well as from the lower investment cost and greater
output per unit invested. For example, efficiency results from the
fact that the caloric heat content can be extracted more completely
from gas than from coal. Also, the same energy fuel can have higher
efficiencies if used in certain applications. Soviet analysts have
pointed out that conversion of industrial furnaces from coal to gas
can result in efficiency gains as high as 15 percent. Meanwhile,
conversion of boilers of large electric stations can result in gains of
2 percent. [13] Relative comparisons of efficiencies of converting from
coal to gas and from coal to oil are reported in Table 6.3.

The decreased growth rate of the Russian economy since 1960
has been of great concern to the Soviet leadership. A major goal of
the Ninth Five-Year Plan is to increase the growth rate through an
improvement in efficiency and productivity. The plan has stressed
the potential advantages inherent in the shift in the energy base. A
priority requirement has been placed on the accelerated development
of the new and highly productive petroleum and gas resources.[14]

A review of the data shows that Soviet energy strategy is dedicated
to the conversion from coal to gas and oil. The multiple public
references and performance data leave no doubt that the Soviets have
the desire and will to develop their resources in Siberia. To provide
the means to implement their energy strategy, Western participation
in this development has been invited, and specific legislative provisions
have been made to permit Western participation.

GAS SUPPLY-AND-DEMAND TRENDS IN THE USSR

Turning from energy strategy to the specific energy component—
natural gas—the ultimate feasibility of the North Star project depends
on the availability of sufficient gas in excess of Soviet domestic needs
to satisfy the projected delivery schedules. (Since the delivery
schedule spans 25 years, it is necessary to take a long-range look at
Soviet demand-and-supply trends.) The past, current, and future
production-and-consumption trends will be analyzed in order to
determine the availability of gas for North Star.

Soviet managers have learned that production and consumption
of gas are a very difficult task to manage. Gas must be managed as
a system. The discovery of new reserves, development of new fields,
and transportation of gas from field to consumers must be carefully
integrated. The Soviet publication Economic Gazette says:

TABLE 6.3

Economic Effect of Converting from Coal to Gas or Oil in the USSR
(using equivalent cost of 1 ton of standard fuel in gas, oil, and coal)

| | Economies on Conversion from Coal per 1 Ton of Standard Fuel | | | | | |
| | Natural Gas | | | Black Oil | | |
Installations	Total	Increased Efficiency	Decreased Investment and Operating Costs	Total	Increased Efficiency	Decreased Investment and Operating Costs
Industrial furnaces	8 – 15	3.0 – 5.0	5.0 – 10.0	5.5 – 8.0	2.5 – 4.0	3.0 – 4.0
Heating boilers	6 – 10	2.5 – 4.0	3.6 – 6.0	4.5 – 7.0	2.0 – 3.5	2.5 – 3.5
Industrial boilers	3.2 – 4.5	0.7 – 1.0	2.5 – 3.5	2.1 – 2.8	0.6 – 0.8	1.5 – 2.0
Boilers of large electric stations	2.0 – 2.3	0.2 – 0.3	1.8 – 2.0	1.6 – 1.8	0.1 – 0.2	1.5 – 1.6
Locomotives	—	—	—	15 – 18	10 – 12	5.6 – 6.0

Source: Mikhail S. L'vov, Resursy Prirodnovo Gaza SSSR [National Gas Resources of the USSR] (Moscow: Nedra, 1969), p. 14.

The successful achievement of the 1973 tasks assigned to the
gas industry is bound up with maximum development of
reserves and increasing the effectiveness of reserves,
rational use of materials, and a clear linking in the work
of all the related aspects of the task.[15]

The entire process requires a managerial sophistication that
creates havoc with the rigidities of a planned economy. Generally
speaking, the goals set over the years in the Five-Year Plans have
not been met for an assortment of reasons including management,
finance, technology, and resource shortages. In recent years, failure
to achieve established goals stems primarily from difficulties in
completing pipeline construction and problems in the development of
required exploration, drilling, processing, and transportation equipment.[16]

The current goals for Soviet gas were established in the Ninth
Five-Year Plan (1971-75):

To increase gas production to 300-320 billion cubic meters
and to increase the proportion of petroleum and gas in the
energy balance to at least 67 percent by 1975.
. . . to accelerate the development of new and highly
productive petroleum and gas deposits.
. . . to increase the recovery of petroleum associated
gas to 80-85 percent of the gas available at the wellhead.
. . . to integrate the processes of associated and natural
gas and to expand production of liquefied gas.
. . . to continue the task of developing a unified gas
supply system to ensure a more efficient flow of gas via
gas pipeline from the new gas fields to the European part
of the country.
. . . to build at least 30,000 kilometers of main gas
pipelines . . . with 1,420 millimeter pipe diameter . . .
introduce new high capacity pumping to transport gas under
75 atmospheres of pressure.[17]

The announced goal of the Ninth Five-Year Plan, to reach a gas
output level of 300 to 320 billion cubic meters (BCM), appears to be
impossible to achieve. The Soviet press, however, continues to hold
to the 320-BCM goal.[18] The schedule of the original Five-Year Plan
has not been achieved, as indicated in Table 6.4.

Although the Soviets have not achieved the desired gas production, they have made impressive gains. They increased gas output
in 1973 by 36 BCM per year, or by 18 percent since 1970. This is

TABLE 6.4

Current Production Goals of Soviet Gas
(in billion cubic meters)

	1970	1971	1972	1973	1974	1975
	198					
Five-Year Plan		211_b 211^b	229 R229 221^b	250 238^b 236^b	280 R257 261^b	320 R285

^{a}achieved output; ^{b}revised goal.

Sources: N. K. Baibakov, State Five-Year Plan for the Development of the USSR National Economy for the Period 1971-75, pt. 2 (Arlington, Va.: Joint Publications Research Service, September 1972), p. 357; Pravda, February 26, 1974; December 15, 1973; Januar 30, 1973; Izvestia, January 25, 1975; and Ekonomicheskaya Gazeta [Economic Gazette], no. 8 (February 1973).

a considerable achievement by any standard. Nonetheless, the plan had established a demand beyond this output level to satisfy domestic and foreign export requirements. Although output was less than that specified in the plan, no reports could be confirmed to indicate that export commitments were not met. The lack of evidence suggests that when shortages are experienced, export requirements enjoy a priority and domestic supply is curtailed.

While the record shows that the Soviets have not achieved the goals set for gas, it should be recognized that these goals were an impossible dream. It was not that the goals were impossible to achieve, but the priority on resource allocation required to achieve the goals made such action unlikely. The cost to other sectors of the economy was too high.[19] Nevertheless, Soviet achievements in exploration, gas field development, and pipeline construction have been remarkable. Gas production grew rapidly from 9 BCM in 1955 to 236 BCM in 1973. Table 6.2 provides statistical comparisons of gas production to other energy forms, but the growth trend of gas production is more dramatically illustrated in Figure 2.

Dramatic changes also have occurred in the geographic location of gas-producing areas of the Soviet Union. Gas production has shifted

FIGURE 2

Soviet Production of Natural Gas, 1950-75
(in billions of cubic meters)

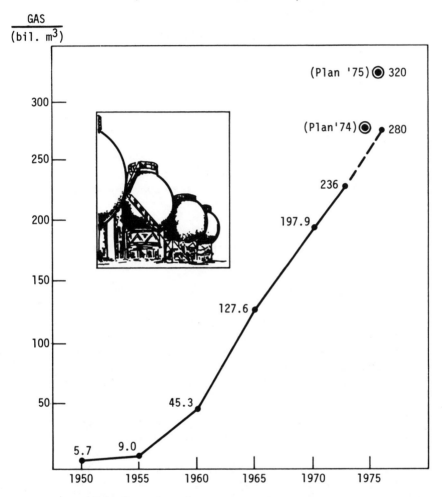

Sources: Mikhail S. L'vov, Resursy Prirodnovo Gaza SSSR
[National Gas Resources of the USSR] (Moscow: Izdatel'stvo "Nedra,"
1969), augmented by various issues of the Soviet statistical yearbook
on the national economy, Narodnoye Khozyajstvo SSSR [National Economy of the USSR].

from almost total production in the European portion of the USSR to
a rapidly increasing production output in Central Asia and Siberia.[20]
The geographic shift in gas-producing areas that has been in progress
since 1960 is reflected in Figure 3.

Over 50 percent of Soviet-proven reserves of natural gas are
located in the West Siberian basin.[21] Four of the five largest gas
fields of the world are located in this basin.[22] West Siberia has the
potential for producing 500 BCM of gas per year, "enough to meet
all the growing requirements of the domestic economy and to expand
foreign economic ties."[23] Thus far, only the Medvezhe deposit has
been brought on-line to supply gas to the Soviet economy. The director
of the Chief Administration for Gas in the Eastern Regions, E. Altunin,
was quoted in March 1973 as saying:

> It is time for us to start work on Urengoi now, this year.
> We cannot succeed without Urengoi. A development pace is
> required that will eventually produce an annual delivery of
> 30 billion cubic meters of gas per year.[24]

Two Soviet journalists visited the West Siberian area in March
1973. They lauded the accomplishments of the workers under severe
hardships, but criticized the slow pace of construction, as follows:

> . . . without Siberian gas, it is impossible to achieve the
> desired changes in the nation's energy balance, and it is
> impossible to achieve the planned increase in the production
> of this valuable fuel during the current Five-Year Plan.
>   . . . Now, when we can no longer get along without
> Tyumen gas, virtually everything must be built in the shortest
> possible time, gas fields, gas pipelines, cities, roads. Sud-
> denly, it is very difficult to cope with everything.[25]

The chairman of the Council of Ministers, A. N. Kosygin,
visited Tyumen on January 13, 1973. The mere visit of such a high-
level official suggests that the gas plan was in trouble and that assets
of this area could ease gas shortages. The Soviet press reported
that workers of the region pledged successful fulfillment of their
portion of the Five-Year Plan. The press indicated that as much as
40 BCM would be produced by 1975 and that the target for the next
Five-Year Plan would be 220 to 240 BCM.[26] This production level,
if achieved, would be equivalent to the total gas production of the
country in 1972.[27]

Available evidence suggests that an adequate potential for gas
supply exists to meet the domestic and export needs of the Soviet

FIGURE 3

Geographic Shift in Gas Producing Areas of the USSR, 1960-74
(in percentage of total production)

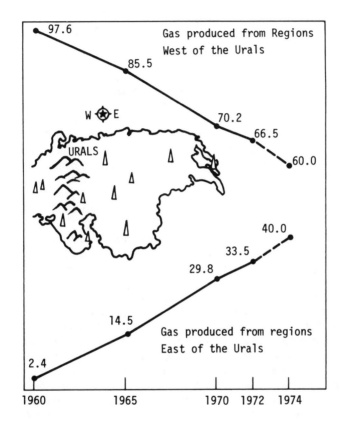

Sources: Mikhail S. L'vov, Resursy Prirodnovo Gaza SSSR
[Natural Gas Resources of the USSR] (Moscow: Izdatel'stvo "Nedra,"
1969), pp. 28-42; and USSR, Staticheskij Otdel, Narodnoye Khozyajstvo
SSSR v. 1972 G. [National Economy of the USSR in 1972] (Moscow,
1973), p. 176.

Union. However, there is a problem of extraction and transportation of the gas within the time schedule established by consumption requirements. A more detailed evaluation of the gas supply and demand trends is necessary to determine whether adequate supplies will be available to support the requirement of North Star for 2.1 BCFD for 25 years.

PROJECTED TRENDS FOR NATURAL GAS DEMAND AND SUPPLY

Any projection of domestic demand for gas must be tenuous at best. The Soviet Union publishes annual statistical yearbooks, which show a decline in the rate of growth of domestic energy from 7.7 percent in the 1950s to about 5.3 percent in the 1965-70 period.[28] The changes in growth rate are a result of policy and a variety of forces including the change in energy balance, technology, and other factors. However, a linear graphic projection of gas consumption from 1960 to 1973 fitted a straight-line trend. It was assumed that this consumption trend would continue at the same rate as in the last decade. On this basis, Soviet domestic demand for gas through 1980 is presented in Table 6.5.

Soviet export demand for natural gas was determined by a summation of the various contractual agreements signed with various East European and West European nations. In addition to the firm contracts signed as of this writing, the Soviets have been negotiating with the United States and Japan for the development and export of natural gas from Western Siberia (North Star) and from Eastern Siberia (Yakutsk). The USSR claims the explored reserves at Yakutsk to be approximately 11 trillion cubic feet (TCF). United States and Japanese technicians estimate that several years of exploratory verification are needed to determine the full potential of the area.[29] Extensive exploratory work, hopefully with U.S. and Japanese financial and technical assistance, will be undertaken so that development can commence.[30] Therefore, export of gas from this area is not considered feasible before the mid-1980s.

On the other hand, the gas reserves of the Urengoi field are proven, and a feasibility study conducted by Tenneco Inc. has determined that the full productive potential of the North Star project could be fully implemented by 1982, if an affirmative decision is reached by 1975.[31] Nevertheless, the export requirement for North Star was not included in total Soviet export demand because no firm contracts have been signed. The total Soviet export demand for gas is summarized in Table 6.6.

TABLE 6.5

Structure of Soviet Natural Gas Demand (Domestic)
(in billion cubic meters)

Category	1961	1967	1971	1975*	1980*
General communal needs	6.5	20.4	27.6	33.0	46.0
Generation of electric power	16.1	40.9	55.6	69.0	96.0
Industry	34.2	91.1	126.7	162.0	210.0
Other uses	2.2	5.0	7.8	10.2	12.7
Total	59.0	157.4	217.7	274.2	364.7

*Estimates for 1975 and 1980 are based on a linear projection of the 1961-71 data trend line.

Sources: Mikhail S. L'vov, Resursy Prirodnovo Gaza SSSR [Natural Gas Resources of the USSR] (Moscow: Nedra, 1969), p. 12; V. V. Strishkov, "The Minerals Industry of the USSR," preprint from the 1971 Bureau of Mines Minerals Yearbook, U.S. Department of the Interior (Washington, D.C.: Government Printing Office, 1971), p. 43; and Aman R. Khan, "Soviet Gas in the Seventies," Pipeline and Gas Journal (October 1972): 25.

Table 6.5 and Table 6.6 establish the estimated domestic and export demand for Soviet gas. Soviet supply or production of gas is estimated on the basis of Soviet-announced plans and the evaluation of those plans by Western experts. Western estimates of Soviet gas production by 1980 vary from 370 to 600 BCM annually.[32] The Soviets themselves have established a production goal of 320 BCM by 1975 and have not set a firm target for 1980, but imply that it will be in the vicinity of 520 BCM.[33] The Soviets are in the third year of their Ninth Five-Year Plan and have revised their interim year goals.[34] Assuming the Soviets will fall short by 15 to 25 percent in meeting their planned gas production goals for 1975 and 1980, it appears probable that the 1975 output will be approximately 280 BCM and the 1980 output 400 BCM. In addition to domestic production, imports of gas from Iran and Afghanistan are expected to increase from 12 BCM

TABLE 6.6

Export Demand for Soviet Natural Gas
(in billion cubic meters)

Country	1970	1971	1972	1973	Est. 1974	Est. 1975	Est. 1980
Communist Countries							
Czechoslovakia	1.2	1.6	1.9	2.4	3.1	3.5	9.0
Poland	1.0	1.5	1.5	1.7	1.8	2.0	3.0
East Germany				0.8	2.8	4.0	4.0
Hungary						1.0	2.0
Bulgaria					1.0	3.0	7.5
Yugoslavia							
Subtotal	2.3	3.1	3.4	4.9	8.7	13.5	25.5
Non-Communist Countries							
West Germany				0.3	1.0	2.0	7.0
Italy						4.0	6.0
Austria*	1.0	1.4	1.6	1.6	2.0	2.0	3.0
France					2.0		2.5
Finland					0.5	1.0	1.4
U.S. (North Star)							(?)
Subtotal	1.0	1.4	1.6	1.9	5.5	9.0	19.9
Total	3.3	4.5	5.0	6.8	14.2	22.5	45.4

*Shipments commenced in 1968.

Sources: USSR, Ministry of Foreign Trade, Vneshnyaya Torgovlya SSSR za 1970 God; 1972 God [Foreign Trade for 1970; 1972 and 1973] [Moscow: Myezhdunarodnyye Otnoshenya, 1971, 1973], various pages; Robert E. Ebel, "Gas in the Soviet Union," paper presented at the 48th Annual Fall Meeting of the Society of Petroleum Engineers, Las Vegas, Nevada, October 1, 1973, p. 8; V. V. Strishkov, "The Minerals Industry of the USSR," preprint from the 1971 Bureau of Mines Minerals Yearbook, U.S. Department of the Interior (Washington Engineers, Las Vegas, Nevada, October 1, 1973, p. 8; V. V. Strishkov, "The Minerals Industry of the USSR," preprint from the 1971 Bureau of Mines Minerals Yearbook, U.S. Department of the Interior (Washington, D.C.: Government Printing Office, 1971), p. 44; J. Richard Lee, "The Soviet Petroleum Industry: Promises and Problems," in U.S. Congress, Joint Economic Committee, Soviet Economic Prospects for the Seventies, Joint Committee Print (Washington, D.C.: Government Printing Office, 1973), p. 290; American Embassy Helsinki Message 0071, January 11, 1974; and a series of interviews and correspondence with J. Richard Lee and Dr. V. V. Strishkov during the period September 1973 to August 1974.

TABLE 6.7

Trends of Supply and Demand for Soviet Gas (in billion cubic meters)

	1955	1960	1965	1969	1970	1971	1972	1973	Est. 1974	Est. 1975	Est. 1980
Supply											
Extracted gas (natural and associated)	9.0	45.3	127.7	181.1	197.9	212.4	221.0	236.0	257.0	280.0	400.0
Manufactured gas	1.4	1.9	1.7	1.7	1.7	1.7	1.7	1.7	1.7	1.7	1.7
Imported gas	—	—	—	2.0	3.6	8.1	11.0	12.4	13.5	15.0	15.0
Total supply	10.4	47.2	129.7	184.8	203.2	222.2	233.7	249.1	271.2	296.7	416.7
Demand											
Surplus											6.6
Domestic consumption	10.4	47.0	129.3	182.7	199.9	217.7	228.7	242.3	260.3	274.2	364.7
Gas exports to Communist nations	—	0.2	0.4	1.2	2.3	3.1	3.4	4.9	8.7	13.5	25.5
Gas exports to non-Communist nations	—	—	—	0.7	1.0	1.4	1.6	1.9	5.5	9.0	19.9
Total demand	10.4	47.2	129.7	184.4	203.2	222.2	233.7	249.1	271.2	296.7	410.1

Sources: USSR, Central Statistical Directorate, Narodnoye Khozyajstvo SSSR v. 1970 G; 1972 G; 1973 G [National Economy of the USSR, Statistical Yearbooks for 1970 through 1973]; USSR, Ministry of Foreign Trade, Vneshnyaya Torgovliya, 1963–73 [Foreign Trade yearbooks for 1963 through 1973]; N. K. Baybakov, "O Gosudarstvennom Plane na 1974 God" [Concerning the Government Plan for 1974], Pravda, December 13, 1973, p. 3; U.S. Department of the Interior, Bureau of Mines Minerals Yearbook, 1971 and 1972 (Washington, D.C.: Government Printing Office, 1972 and 1973); and interviews and correspondence with J. Richard Lee and Dr. V. V. Strishkov during September 1973 and August 1974.

in 1973 to 15 BCM estimated for 1980.[35] A tabulation of Soviet supply and demand of natural gas is summarized in Table 6.7.

A Soviet increase in gas production to 400 BCM by 1980 would meet domestic gas demands equal to the past historic growth trend for domestic consumption. In addition, it would satisfy the export demand from firm contracts signed with East and West European nations and would provide an excess of some 6.6 BCM, or approximately 233 billion cubic feet (BCF) of gas. This surplus could provide about 0.6 BCF per day for support of North Star by 1980.

The achievement of a 400-BCM production is not unreasonable, by comparison to past performance. During the current Five-Year Plan, the Soviet Union planned to increase gas production by approximately 50 percent, from 198 to 300 or 320 BCM. Results now indicate that they will probably achieve an output of 280 BCM by 1975, or a five-year growth of approximately 42 percent. In order to achieve a 400-BCM output by 1980, a growth of almost 43 percent is required. If past history is a guide, the Soviets will probably set their output goal at approximately 450 BCM. However, achievement of 400 BCM is more likely. Therefore, gas resources appear to be available to support North Star without significant sacrifice of domestic use.

SUMMARY

In this chapter, Soviet energy and natural gas strategy were analyzed to determine whether adequate quantities of natural gas would be available to meet the requirements of the North Star project. The findings are

1. Soviet energy strategy has shifted from coal to oil and gas as the energy base.

2. The Soviet Union is recognized as having the largest resources of natural gas in the world.

3. Production of natural gas has increased dramatically from 9 to 236 BCM in the period from 1955 to 1973. Consumption has increased on a comparable basis as a result of the shift in energy base.

NOTES

1. Nicolas Spulber, Soviet Strategy for Economic Growth (Bloomington: Indiana University Press, 1964), p. 75.

2. Eugene Zaleski, "Planning for Industrial Growth," in Development of the Soviet Economy, ed. V. Treml (New York: Frederick A. Praeger, 1968), pp. 55-56.

3. Robert W. Campbell, The Economics of Soviet Gas and Oil (Baltimore: Johns Hopkins Press, 1968), pp. 2-3.

4. N. I. Buyalov, et al., Metodika Otsenki Prognoznykh Zapasov Nefti i Gaza [Methods for Evaluating Predicted Reserves for Gas and Oil] (Leningrad: Gox Top Tekh Izdat, 1962), pp. 5-11.

5. A. P. Agishev, V. G. Vasiliev, and Y. M. Vasiliev, "Principal Gas-Bearing Areas of the Soviet Union," paper presented at the 12th World Gas Conference, Paris, April 1973, pp. 2-5.

6. Michael T. Halbouty, Geology of Giant Petroleum Fields (Tulsa: American Association of Petroleum Geologists, November 1970), pp. 508-09.

7. Campbell, op. cit., pp. 1-6.

8. Ibid., pp. 10-15.

9. Ibid., pp. 8-9.

10. Ibid., pp. 9-10.

11. USSR Central Statistical Directorate, Narodnoye Khozyajstvo v. 1970 g. [National Economy of the USSR in 1970] (Moscow: Statistika, 1971), p. 185. It should be noted that the 1972 statistical yearbook, Narodnoye Khozyajstvo v. 1972 g. [National Economy of the USSR in 1972] (Moscow: Statistika, 1973) omits the tabulation for manufactured gas. Trend projections reveal that manufactured gas would have constituted less than 1 percent of the total gas output in 1972. It is assumed that gas continued to be manufactured but that tabulations are dropped because of the relatively small quantity produced.

12. "Direktivy XXIV S'yezda KPSS po Pyatiletnemu Planu Razvitiya Narodnovo Khozyajstva SSSR na 1971-1975 Gody" [Directives of the 24th Congress of C.P.S.U. Concerning the 5-Year Plan for the Development of the National Economy of the USSR during the Period 1971-75], Pravda, February 14, 1971, p. 2.

13. Mikhail S. L'vov, Resursy Prirodnovo Gaza SSSR [Natural Gas Resources of the USSR] (Moscow: Nedra, 1969), pp. 13-14.

14. "Direktivy XXIV S'yezda KPSS," op. cit., p. 2.

15. "Gazovaya Promyshlennost" [Gas Industry], Ekonomicheskaya Gazeta [The Economic Gazette], no. 8 (February 1973): 2.

16. Ibid., and Campbell, op. cit., p. 198; Robert E. Ebel, "Russia Falls Short of Pipe Line Goals," Pipe Line Industry, November 1973; and V. Sukhanov and E. Shatokhin, "Tochka Otchyeta: Zapadno-Sibirskij Kompleks: Opyt i Problemy" [Bench Mark: West-Siberian Complex: Experience and Problems], Izvestia, March 20, 1972, p. 2.

17. "Direktivy XXIV S'yezda KPSS," op. cit., p. 2.

18. "Gazovaya Promyshlennost," op. cit., p. 2.

19. John P. Hardt, "West Siberia: The Quest for Energy," Problems of Communism (May-June 1973): 28-29.

20. L'vov, op. cit., pp. 25-32; and USSR Central Statistical Directorate, op. cit., p. 207.

21. Agishev, Vasiliev, and Vasiliev, op. cit., pp. 4-5.

22. Halbouty, op. cit., pp. 508-09.

23. V. Sukhanov and E. Shatokhin, "Bolshoi Gaz Zapolyr'ya" [Huge Gas of the Arctic], Izvestia, March 23, 1973, p. 3.

24. V. Sukhanov and E. Shatokhin, "Rabochiye Trassy Severa" [Workers' Tracks in the North], Izvestia, March 21, 1973, p. 3.

25. Sukhanov and Shatokhin, "Bolshoi Gaz Zapolvar'ya," op. cit., p. 3.

26. Ibid.; and "Namyecheny Vysokiye Rubyezhi" [High Goals are Set], Pravda, January 14, 1973, p. 1.

27. USSR Central Statistical Directorate, op. cit., p. 207.

28. Robert W. Campbell, "Some Issues in Soviet Energy Policy for the Seventies," in U.S. Congress, Joint Economic Committee, Soviet Economic Prospects for the Seventies: A Compendium of Papers Submitted to the Joint Economic Committee, ed. John P. Hardt, Joint Committee Print (Washington, D.C.: Government Printing Office, 1973), pp. 49-50.

29. Agishev, Vasiliev, and Vasiliev, op. cit., p. 5; and U.S. Central Intelligence Agency, The Soviet Economy in 1973: Performance Plans and Implications, A(ER) 74-62 (Washington, D.C.: Central Intelligence Agency, July 1974), p. 25.

30. Raymond J. Albright, Siberian Energy for Japan and the United States, case study for the Senior Seminar in Foreign Policy, Department of State, 1972-73; and Don Oberdorfer, "Soviets Unblock Japan's Role in Siberian Oil and Gas," Washington Post, March 10, 1974, p. A-20.

31. Jack H. Ray, interviews and correspondence, September 1973 to July 1974.

32. J. V. Licence, "Siberia in the Context of World Natural Gas Supplies," paper prepared for NATO Round Table Meeting, Brussels, January 30, 1974, p. 8; and V. V. Strishkov, "The Minerals Industry of the USSR," preprint from the 1971 Bureau of Mines Minerals Yearbook, U.S. Department of the Interior (Washington, D.C.: Government Printing Office, 1971), p. 20.

33. "Direktivy XXIV S'yezda KPSS," op. cit., p. 2.

34. "Gazovaya Promyshlennost," op. cit., p. 2.

35. Interview with V. V. Strishkov at the U.S. Department of the Interior, Washington, D.C., August 8, 1974.

7

SOVIET
TECHNOLOGY

The technology level in many sectors of the Soviet economy is below Western levels, and the 1974 plan is dedicated to the improvement of economic performance through technological upgrading. The plan contains provisions for procurement of foreign equipment in the hope of improving economic performance. The gross national product (GNP) of the Soviet Union is about half that of the United States, while its labor force is 40 percent greater, and the rate of capital investment is double that of the United States.[1] The reasons for this discrepancy in performance are many and include technology lag, inefficient management practices, a rigid planning system, and an emphasis on quantity as opposed to quality output.[2]

In the broad view of technology, both Western and Soviet experts agree that Soviet technology lags considerably behind the West, except in certain military and space sectors. In May 1958, Premier Khrushchev complained that

. . . the entire procession of new techniques and products generated by the welter of discoveries made within the modern chemical industry had literally bypassed the Soviet economy, while its leaders were preoccupied with the pursuit of the will-of-the-wisp of "technical and economic independence," that is, with protection of the domestic economy against possible negative influences from the outside world.[3]

On the basis of an analysis of the introduction of key technological innovations into the Soviet economy, it is estimated that in 1962 Soviet technology was 25 years behind that of the United States.[4] The technology lag varies by sectors, and the Soviet economy has been described as having a first-rate military sector, a second-rate industrial sector, and a third-rate consumer sector.[5]

Productivity is often used as an indicator of the technology lag, for it reflects the output achieved per unit of input. The Soviet economy is characterized by the highest rate of capital investment in the world. In 1971, the share of GNP allocated to capital investment was 31 percent, compared to 17 percent in the United States. However, labor productivity in the two countries has grown at approximately the same rate since 1950. Labor productivity in Soviet industry was about 40 percent of levels in the United States in 1955 and 1971. Similarly, productivity of Soviet agriculture was approximately 11 percent of levels in the United States in both periods.[6]

Premier Aleksei Kosygin complained in 1965 that "the pattern of production of machinery and equipment being turned out by the many branches [of Soviet industry] does not conform to modern standards." In 1970, Soviet scientists Sakharov, Turchin, and Medvedev criticized Soviet backwardness in energy, chemistry, and computers.[7]

In an effort to partially overcome the technology gap, the Soviet Union has been importing equipment, including turnkey plants, from the West. During the decade of the 1960s, the average annual balance-of-payments deficit was approximately $250 million per year.[8]

The 1974 plan for the Soviet economy recognized that foreign sources made it possible for the national economy "to obtain modern machinery and equipment, as well as other goods, not manufactured in the USSR," and the plan included provisions for continued procurement of foreign equipment.[9]

TRANSFER OF TECHNOLOGY

Transfer of technology can be effected through several channels. In order to understand the aspects of technology that are involved in the North Star project, a discussion will be presented on the different concepts of technology transfer, the state of Soviet technology, and the specific technological needs of the Soviet gas industry. The impact of technology transfer on the Soviet gas industry through North Star will be evaluated.

In simplest terms, the sale of an advanced tool or machine from one nation to another can be viewed as technology transfer. A higher order of transfer occurs when the factory that manufactures the tool is sold. Another aspect of technology transfer involves the techniques or managerial "know-how" needed to make the employment of that tool effective in the importing economy. The most sophisticated transfer would be the transplant of the creative genius which invented the tool to the importing economy.[10]

Another approach to transfer of technology concerns the effects
of such transfer on the firm or nation exporting the technology. The
transfer can be considered the creation of a competitor. The natural
consequence of a new competitor in the market is the loss of market
share. The resultant loss of sales causes a reduction in resources
available for research and development and the eventual loss of the
leadership role.[11] In extreme cases, transfer of technology can be
considered a threat to survival. If all technology is considered to
have military as well as commercial application, technology transfer
can be viewed as "national suicide."[12]

There are implied assumptions in the above concepts of technology
transfer which tend to be accepted as fact, but which lack a foundation
in truth. These assumptions are

1. No interface problems exist.
2. The importing economy can join the technological learning curve
   and can continue to advance technologically.
3. The market demand for the product produced is fixed.
4. The United States could block technology transfer to the Soviets.

The first implied assumption minimizes or ignores the problems
of environmental interface involved when a new tool or advanced
machine is introduced into an economy. It assumes that the tool will
work as efficiently in the Soviet economy as it did in the economy of
the United States. Furthermore, the transfer of technology among the
various sectors of the Soviet economy is very difficult, and the military
sector operates in a state of technological quarantine from the civilian
sector.

The segmentation of the Soviet economy, and particularly the
aversion to technology transfer between the civilian and military
sectors of the economy, was recently described as follows:

> . . . the military and civilian sectors of the Soviet economy
> are considered separate and distinct:
>   The two sectors operate on different technical levels,
> and according to quite different rules, and with a consider-
> able secrecy barrier. It is clear that the leaders had a very
> difficult time trying to transfer to the civilian sector the
> managerial techniques, the innovative behavior, and high
> quality that seem evident in the military and space sector.[13]

An example of the problems of technology transfer within the
Soviet economy is reflected in the Russian experience with computers.
The demonstrated Soviet capability to land instrument packages on the

moon and return them to earth requires a sophisticated computer capability to handle the telemetry data. The domestic economy is placing high hopes on the computer for improving its productivity and efficiency. But it has had difficulty in manufacturing and employing the computer effectively and is anxious to import Western technology to help solve its problems. Robert Campbell stated recently:

> . . . It has sometimes been said, only partly in jest, that when one thinks of the problems the Russians have in getting computers maintained, in fitting them into their procedures and systems, and providing the software and modeling support, the large-scale importation of Western computers would do more to set back the progress of Soviet planning and management than anything else one could imagine.[14]

The American planners of North Star realize the interface problem. They believe that the only sure way to provide a reliable source of gas would be to control the system all the way from the Soviet gas field to the American tap.[15] The project is planned to operate independently and to avoid interface and possible interruption resulting from a mixed Soviet-U.S. equipped project.

A second implied assumption is that once technology is transferred, the recipient will step into the learning curve at that stage of development and will continue to advance into higher stages of technology. A historical experience suggests a fallacy in such reasoning. A transfer of U.S. technology was effected in the 1930s into the Soviet auto industry by

> . . . the 1930 Ford Motor Company agreement to build a completely new integrated plant for mass production of the Model A, the 2.5-ton Ford truck, and buses using Ford patents, specifications, and manufacturing methods. The plant was erected by Albert Kahn, the builder of River Rouge, and so enabled the Soviets to duplicate the immense advantages of American automobile engineering within a few years of inception in the United States.[16]

Although the Ford Company made its latest technical processes and trade secrets available to the Russians and trained Russian engineers and skilled workmen to operate the plant, the automobile technology imported failed to advance and keep up with world advances. By 1969, the Soviets signed an agreement with the Italians to build the Fiat automobile plant in order to bring up to date their obsolescent technology. This action suggests that transfer of technology involves

considerably more than the sale of equipment and the training of
personnel in management and operating techniques. Rapidly changing
technology requires continued contact with outside sources. Efficient
transfer of technology requires an opening of Soviet society, which
the ideology condemns.

A third implied assumption is that market size for a product is
relatively fixed, and the transfer of technology will create a competitive producer who will enter the market and take over a share of the
existing market. The reason for North Star is inadequate supply.
The American gas companies are unable to meet the market demand
for gas. Development of the Siberian resources would alleviate the
world gas shortage and, hopefully, reduce demand pressures. It is in
this area that the transfer of technology through North Star differs
radically from transfers in other sectors. North Star embodies the
classical principle of comparative advantage. It is barter trade in
which each nation trades a commodity in which it enjoys a comparative
advantage. The commodities come, however, from sectors that are
physically unrelated and economically separated. Usually goods are
traded for goods, or technology is traded for technology. In the North
Star project, Soviet natural gas is bartered for equipment and technology, under special financing arrangements. A competitor is not
created. There is no loss of market share. More important, the sale
of equipment and managerial fees provide the United States with
resources for future research and development to keep the U.S.
industry healthy and growing.

A fourth implied assumption is that the United States could block
the transfer of technology to the Russians. Gas technology is widely
diffused throughout the world, and other sources of gas technology
are available to them. The gas technology involved in North Star
embraces a broad spectrum of items, including gas gathering, pipeline technology, and liquefied natural gas (LNG) technology. The LNG
technology will be analyzed first because it is the most frequently
mentioned item and is the newest technology from the standpoint of
development.

TRANSFER OF LNG TECHNOLOGY

The Soviets are not considered to possess LNG technology even
though the United States is reported to have supplied them with an
LNG plant, located outside Moscow, under Lend-Lease.[17] In the
North Star proposal, the transfer of LNG technology would be
accomplished through the sale of a liquefaction plant, storage

TABLE 7.1

International LNG Agreements—Operating, Planned, Proposed—as of December 1973

From	To	Companies Involved	Initial Deliveries
Algeria	Gr. Britain (Canvey I.)	CAMEL, Conch International	1964
Algeria	France (Le Havre)	Gaz de France, CAMEL	1965
Alaska	Japan	Phillips-Marathon, Tokyo Gas Tokyo Electric	1969
Libya	Italy (La Spezia)	Esso Std. Libya, SNAM	1971
Libya	Spain (Barcelona)	Esso Std. Libya, Gas Natural S.A.	1971
Algeria	France (Fos)	Sonatrach, Gaz de France, Somalgaz	1972
Brunei	Japan	Brunei Shell Petr. Co.	1972
		Brunei LNG Ltd.	1973
		Coldgas Trading Co.	1974
		Tokyo Electric Co.	
		Osaka Gas Co.	
Algeria	U.S.	Sonatrach, Alocean Ltd.	1971
		Distrigas	1974
			1975
Algeria	U.S.	Sonatrach, El Paso Natural Gas Co.	1975
		Columbia LNG Corp.	
		Consolidated System LNG Co.	
		Southern Energy	
Algeria	U.S.	Sonatrach, El Paso, Transco Energy	1976
Algeria	U.S.	Sonatrach, El Paso	1976
Algeria	U.S.	Sonatrach, Eascogas LNG Inc.	1976
		(Pub. Sv. Gas & Elec. of N.J.	1977
		Algonquin LNG Co.)	
Algeria	U.S.	Panhandle Eastern Pipe Line Co.	1980
Trinidad	U.S.	Amoco Trinidad Oil Co., Peoples	1976
		Gas (Natural Gas Pipeline Co.	1977
		of America)	1978
Algeria	Spain	Sonatrach, Gas Natural S. A.	1976
Algeria	Belgium	Sonatrach, Distrigaz	1977
Algeria	West Germany	Sonatrach, Bayerische Ferngaz, Gasversorgung Suddeutschland	1977
Indonesia	U.S.	Pertamina, Pacific Lighting Corp.	1978
Alaska	U.S. (Los Angeles)	Pacific Alaska LNG Co.	1976

From	To	Companies Involved	Initial Deliveries
Alaska	U.S.	Northwest Natural Gas Co.	1976
Ecuador	U.S.	Pacific Lighting Ada Exploration	1977
Venezuela	U.S.	Venezuela Ministry of Mines, CVP (Punta de Palmas site) unknown U.S. firm	1977
Abu Dhabi	Japan	Abu Dhabi Marine, Bridgestone Liquefaction Gas Co., Mitsui	1977
Australia (Bonaparte Gulf)	Japan	Tokyo Gas	?
Australia (Dongara Gulf)	Japan	Tokyo Gas	?
Australia (Palm Valley)	Japan & U.S.	Magellan Petr. Austr. Ltd.	1976
Australia (Northwest Shelf)	Japan	BOC of Australia, Mitsui C. Itoh & Co., Sumitomo Trading	1976
Iran	U.S.	Unknown	1977
Iran	Japan	National Iranian Gas Co., Fuji Oil, Marubeni	1977
Iran	Japan	C. Itoh & Co.; NIGC; Osaka Gas; Kansai Electric	1976
Iran	Japan	Phillips, Mitsui	?
Sarawak	Japan	Royal Dutch Shell, Tokyo Gas Tokyo Electric, others	1977
Nigeria	U.S.	Phillips	1976
Nigeria	U.S.	Shell Intl. Gas; BP Development of Nigeria	?
Nigeria	U.S.	Gulf Oil, unknown	?
New Guinea	Japan	Unknown	?
Chile	Chile	Empresa Nacional del Petroleo	?
Iraq	Unknown	Iraq National Oil Co.	1977
USSR	U.S. East Coast	USSR, Texas Eastern, Tenneco, Brown & Root, Inc.	1980, later
USSR	U.S. West Coast	Occidental Petr., El Paso, USSR	1980, later

Source: Dean Hale, "International LNG Movements Continue to Expand," Pipeline and Gas Journal (December 1973): 61.

MAP 2

LNG Projects in Operation, January 1, 1974
(in millions of cubic feet per day)

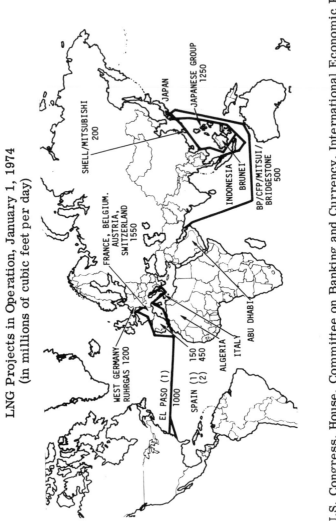

Sources: U.S. Congress, House, Committee on Banking and Currency, International Economic Policy, Hearings before the Subcommittee on International Trade on H.R. 774, H.R. 13838, H.R. 13839, and H.R. 13840, 93rd Cong., 2nd sess., 1974, p. 260; and Dean Hale, "International LNG Movements Continue to Expand," Pipeline and Gas Journal (December 1973): 61.

MAP 3

LNG Projects under Construction, January 1, 1974
(in millions of cubic feet per day)

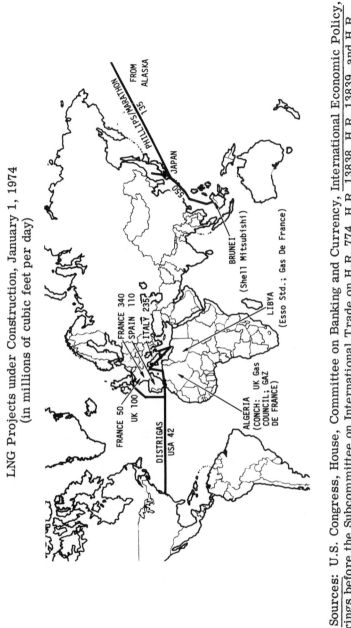

Sources: U.S. Congress, House, Committee on Banking and Currency, International Economic Policy, Hearings before the Subcommittee on International Trade on H.R. 774, H.R. 13838, H.R. 13839, and H.R. 13840, 93rd Cong., 2nd sess., 1974, pp. 259-61; and Dean Hale, "International LNG Movements Continue to Expand," Pipeline and Gas Journal (December 1973): 61.

TABLE 7.2

Summary of Long-Term LNG Import Projects (as of November 9, 1973)

Exporting Country	Status	Year of Initial Delivery	Contract Term (Years)	Companies Involved	Daily Gas Volume (million cubic feet)	Number of Ships and Country of Construction	Capacity of Ships (cubic meters)
Algeria	1	1971	20	SONATRACH			
				Alocean Ltd.		1, France	50,000
				Distrigas Corp.	300.0	1, France	120,000
Algeria	1	1976	25	SONATRACH			
				El Paso Natural Gas Co. (Algerian Subsidiary)		3, France	125,000
				Columbia LNG Corp.	300.0	6, U.S.	125,000
				Consolidated System LNG Co.	350.0		
				Southern Energy Co.	350.0		
Total					1000.0		
Algeria	1	1975	22	SONATRACH			
				EASCOGAS LNG, Inc.		3, U.S.	125,000
				(Staten Island, N.Y., Terminal)	424.0	2, (NA)	125,000
				(Providence, R. I., Terminal)	228.0		
Total					652.0		
Algeria	1	1976	20	SONATRACH			
				El Paso Natural Gas Co. (Algerian Subsidiary)		6, U.S.	125,000
				El Paso Eastern Co.	373.3	3, (NA)	125,000
				Consolidated System LNG Co.	150.0		
				Southern Energy Co.	100.0		
				Transco Energy Co.	373.3		
Total					996.6		
Algeria	1	1979	20	SONATRACH			
				Panhandle Eastern Pipe Line Co.	450.0	5, (NA)	125,000

Country	Status	Year		Company			
Indonesia North Sumatra	1	1978	20	Pertamina-Mobil Pacific Lighting International SA	550.0	8, (NA)	130,000
U.S. (Alaska)	2	1976	—	Phillips-Marathon Northwest Natural Gas Co.	40.0	1, U.S.	
U.S. (Alaska)	2	1977	20	Pacific Alaska LNG Co. (Increasing to 400.0 if gas available)	200.0	1, U.S. (1, U.S.)	130,000 (130,000)
Trinidad	2	1980	20	Ministry of Petroleum and Mines Amoco Trinidad Oil Co. Natural Gas Pipeline Co. of America	400.0	2, U.S.	125,000
Venezuela	2	1977	—	CVP Ministry of Mines and Hydrocarbons (U.S. purchase to be selected)	700.0	4, (NA)	125,000
USSR	3	1980	25	Brown and Root, Inc. Tenneco Inc. Texas Eastern Transmission Corp.	1000.0 1000.0 2000.0	20, (NA)	120,000
Total							
USSR	3	1980	25	Occidental Petroleum Corp. El Paso Natural Gas Co.	1000.0	10, (NA)	125,000
Nigeria	3	1978	—	Ministry of Mines and Power (Oil company and U.S. purchaser to be announced)	1100.0	9, (NA)	120,000
U.S. (Alaska)	3	1978	—	El Paso Alaska	1200.0	6, U.S.	125,000

Note: Status: 1, FPC application filed for import into the United States; 2, prospective applications with FPC; 3, possible application with FPC.

Source: Gas Supply Review, December 1973, p. 61.

facilities, and perhaps LNG tankers. Conceptually, it would be the
lowest order of technology transfer in that it would "sell" tools and
equipment and would not transfer the ability to build LNG equipment
or the capability to design it. The state of LNG technology in the
world and the U.S. position in that environment are discussed below.

The first LNG deliveries in commercial quantities were made
from Algeria to Canvey Island, Great Britain, in 1964. A series of
LNG deliveries was made to Le Havre, France (1965), Tokyo, Japan
(1969), La Spezia, Italy (1971), and Barcelona, Spain (1971) before
the first deliveries were made to the United States--to Boston, in
1971. A total of 40 LNG agreements, which cover shipping of LNG
in all areas of the globe, are in various stages of implementation,
from operating to planned,[18] as shown in Table 7.1. The worldwide
LNG projects that are in operation are portrayed in Map 2, and
those that are well along in the development stage are portrayed in
Map 3.

Out of the total worldwide, long-term LNG projects, a total of
14 would deliver gas to the United States. Only one U.S. project is
currently in operation, but six projects have applications on file
with the Federal Power Commission (FPC).[19] Table 7.2 contains
details concerning the projects.

In support of the various LNG projects, 14 LNG tankers are
operating, and 49 are under construction. The status of the present
LNG tanker fleet is shown in Table 7.3.

TABLE 7.3

Status of the World LNG Tanker Fleet as of January 1, 1974
(capacity 25,000 to 130,000 cubic meters)

Status	Number of Ships
Operating	14
Under construction	49
Proposed	87
Total	150

Source: Compiled from testimony of Jack Ray, in U.S. Congress,
House, Committee on Banking and Currency, International Economic
Policy, hearings before the Subcommittee on International Trade,
93rd Cong., 2nd sess., 1974, pp. 252-57.

Univarsitas
BIBLIOTHECA 99
Ottaviensis

The evidence demonstrates that LNG technology, the ability to liquefy natural gas and ship it by sea, is well developed and is available throughout the world. Therefore, the United States cannot prevent Soviet access to it.

TECHNOLOGY REQUIRED FOR ARCTIC OPERATIONS

The precise definition of gas technology, which would be transferred under the North Star project, is difficult to establish in terms of gas industry infrastructure, such as metallurgy, casings compressors, and pipelines. For example, the United States does not have the capability to manufacture the 48-inch pipe which is required for the pipeline from Urengoi field to Petsamo. If the North Star project is approved, the United States will develop a capability to manufacture 48-inch pipe. This capability would not only aid North Star but also would place the United States in a competitive position to supply large-diameter pipe for other long-range transmission lines now in the planning stages. The fact that pipe for the Alaska pipeline was bought in Japan is perhaps the best evidence that the United States is not currently competitive in this area.[20]

The increased demands and reduced supply induced by the depletion of operating gas fields in the USSR faster than anticipated have placed increased emphasis on the resources situated in Siberia. The key to extracting these resources is technology. A recent Soviet book concludes with the following comment:

> Creation of the technology needed for Arctic development is of enormous importance in freeing the riches of Southern Siberia. The building of a permanent and active means of transportation (railroad and highways) in the territories of the potential gas-producing regions is vital, because without these facilities, it is impossible to achieve the contemplated levels of development.[21]

The hostile Arctic environment, isolated Siberian locations, and high construction costs could be expected, in themselves, to subvert Soviet attempts to achieve their gas goals in Siberia. But the Soviets have compounded their difficulties by introducing large-diameter drilling wells, by the employment of new turbodrill technology, and by accelerating the time schedule of development.[22]

In attempting to master Siberia, the Soviets seem to have delayed too long trying to decide how to approach the many problems involved

in Siberian development. The shift in energy balance away from coal
caused a sharply increasing demand for gas. Simultaneously, accelerated depletion of older gas fields caused supply shortages. In an
almost frantic reaction to the shortages, development resources were
rushed to Siberia, but they were not well coordinated or managed.
Delays and loss from inefficiency resulted. The Soviet press contained
reports concerning items such as the nonavailability of compressors
and equipment, delay in pipe delivery, and conflict over responsibility
for construction and allocation of costs when employing the new modular concept for construction in the Tyumen region.[23] The extent of
confusion is reflected in these words:

> The Ministry of Oil and Gas Construction should take decisive
> measures to coordinate the work of construction and assembly
> organizations. The most important missed objectives of the
> current [1972] plan appears to be construction at
> Kretishchensk, Naip, Shatlyk, Vuktyl, Mevezh'ev, Urtabulak
> fields, and construction of gas pipelines and compressor
> stations in Central Asia, and Central and Northern Tyumen
> Oblast', Urals (Ukhta), Torzhok (Orenburg), and Kuibyshev,
> expansion of the Kiev system of gas pipelines.[24]

Operations in Siberia require not only a mastery of new technology but also the availability of high-quality equipment. The USSR
lacks sophisticated geophysical tools, such as airborne seismic
equipment and computerized field units, which are standard equipment
in the West. Inefficiency in exploration is compounded by poor-quality
drilling equipment. Inferior-quality metallurgy is reflected in
defective pipe, drill casings, and drill bits. Underpowered mud pumps
compound drilling inefficiency. Heavy reliance on the turbodrill,
which is grossly inefficient when drilling below 8,000 feet, causes
rising costs and reduced drilling rates.[25] Richard Lee says:

> . . . Soviet tubular steel goods—drill pipe, casing, tubing,
> pump rods, linepipe, and so on—are known to contain flaws
> and impurities that make them very brittle and subject to
> breaking on impact at low temperatures. Improper well
> design, poor drilling fluid technology, and the lack of suit-
> able blow-out preventers have been major causes of drilling
> accidents in the area to date. Failure to insulate the drilling
> column properly has resulted in the loss of some gas wells,
> and formation of hydrates in the wells has caused break-
> downs of equipment in service. The laying of gas pipelines
> in the permafrost has been accomplished but not without

serious problems of maintenance and operation. Valves
and fittings have cracked; temperature variations and ground
heave have caused sections of pipeline to slip off of polings,
and some sections laid on the ground sank out of sight because the surface proved unstable. The history of efforts
thus far indicate the permafrost problems have been the
most serious obstacle to northern gas field development,
a fact acknowledged by Premier Kosygin in 1972.[26]

The development strategy used in Soviet Siberia appears to be
based on high capital intensity and light manning. It hopes to conserve investment funds, as well as material and labor resources, by
wide-diameter wells, automatic operations, and large-diameter pipelines. In planning pipeline systems in the United States, engineers
generally recommend building the maximum-size pipeline that can
be justified to meet anticipated needs. In pipeline operation there is
a direct cost-benefit tradeoff. As pipe diameter is increased, costs
increase linearly, but capacity increases exponentially. Therefore,
significant cost savings are possible through use of wide diameters.[27]

Conceptually, the adoption of wide-diameter wells in Siberia
involves a new technology of drawing from several horizons in a
given well at once and slant drilling so that the wellheads for a large
number of wells can be clustered on an artificial island.[28] This system has been used on two fields in Tyumen Oblast—Medvezhe and
Urengoi. Medvezhe is presently delivering gas to the Ural basin via
the Punga-Medvezhe pipeline. Urengoi has been drilled, and the
contemplated flow appears feasible on a test basis.[29]

A Soviet account of these developments stated that the major
impetus for the wide-well approach was a requirement to satisfy the
sharply rising demand for gas by the Soviet economy. Old methods
of gas extraction were inadequate. A new way had to be found to
increase extraction from the new Arctic fields to 75 to 100 billion
cubic meters (BCM) annually, and quickly. The first exploratory
well (number 22) was drilled at Urengoi, using a 219-millimeter
shaft in place of the customary 114-millimeter size. A productive
gas horizon was found at a depth of 1,043 to 1,253 meters. Perfora-
tion was accomplished through a series of drillings using ten openings
of 1 meter (in all, 167 meters were opened). Thereafter, a long
period of experimentation followed, in which results and costs were
tabulated.[30]

A second well was sunk at Medvezhe using a 245-millimeter
shaft. The wide-well approach resulted in a reduction in the total
number of wells required by a factor of four. The first wide-well
extraction system in the world is currently operating at Medvezhe
and has a capacity to provide 70 to 75 BCM per year.[31]

The efficiency of wide-diameter pipe is particularly apparent in the transportation of gas. Production and consumption areas in the Soviet Union are separated by long distances extending to 3,000 kilometers and more over inhospitable terrain. In order to transport the large volumes of gas required at a reasonable cost, wide-diameter pipelines operating under high pressure are required. Published Soviet data confirm that studies were made to determine the principal advantages of wide-diameter technology and that the "attractive economies in metal were particularly desired."[32]

In comparing the relative costs of construction and performance capacity of 1,020-millimeter and 325-millimeter pipeline, Soviet researchers found that unit metal input is decreased by a factor of almost six, development expenses by a factor of five, and capital expenditures by a factor of four. However, pipeline capacity is increased by a factor of 23.[33]

The relative advantages of pipeline diameter, operating pressure, and location of compressor stations on the capacity of a pipeline are reflected in Table. 7.4.

Current and future plans for the Soviet gas transportation system are based on the construction of wide-diameter pipelines of 1,020-, 1,220-, and 1,420-millimeter size. The cost in terms of metal and capital investment relative to the size of pipeline diameter and its effect on throughput capacity are provided in Table 7.5.

TABLE 7.4

Flow Capacity of Soviet Pipelines (in billion cubic meters)

Distance Between Compressor Stations (kilometers)	Diameter of Gas Pipeline (millimeters)		
	1020	1220	1420
Operating Pressure 55 Atmospheres			
100	10.0	16.0	23.8
150	8.2	13.1	19.4
200	7.1	11.3	16.8
Operating Pressure 75 Atmospheres			
100	13.8	22.1	32.8
150	11.3	18.0	26.8
200	9.8	15.6	23.2

Source: Dmitrij V. Belorusov, Osvoyeniye Neftyanykh Mestorozhdenij Zapadnoj Sibiri [Mastering the Oil Deposits in Western Siberia] (Moscow: Nedra, 1972), p. 136.

TABLE 7.5

Specific Planning Factors for Soviet Gas Pipelines
(Index: 1,020-millimeter diameter gas pipeline equal to 1)

Specific Factor	Operating Pressure (atmospheres)	Pipeline Diameter (millimeters)		
		1,020	1,220	1,420
Metal input	55	1.00	0.89	0.824
	73	1.32	1.18	1.08
Capital invest-ment	55	1.00	0.82	0.72
	73	1.32	1.03	0.95
Flow capacity	55	1.00	1.60	2.37
	73	1.32	2.12	3.15

Source: Dmitrij V. Belorusov, Osvoyeniye Neftyanykh
Mestorozhdenij Zapadnoj Sibiri [Mastering the Oil Deposits in
Western Siberia] (Moscow: Nedra, 1972), p. 136.

The average diameter of pipe in the Soviet gas pipeline system at
the beginning of the 1960s was approximately 50 percent larger than
the average diameter of the pipe in the U.S. system. This is an
advantage which accrued to the Soviet Union because its capital base
was built at a later stage of the developing gas technology, and
advantage was taken of the world knowledge in gas technology. The
existing Soviet system, however, has not been operated as efficiently
as the older U.S. system. Low throughput performance relative to
capacity has been a continuing problem for the Soviets, due largely
to compressor difficulties. Conflicts have arisen about compressor
design, and delays have been encountered in installing compressor
stations during the early stages of system development.[34]
The 1,020-millimeter and 1,420-millimeter diameter pipelines
programmed in the Ninth Five-Year Plan will require very large,
gas-turbine-powered compressors of up to 25,000-kilowatt capacity.
In the past decade, the Soviet gas industry has had considerable
difficulty handling much smaller, 6,000- to 10,000-kilowatt compressors.[35] This problem continues to plague Soviet managers who are
attempting to develop the energy resources in the Arctic. According
to a Russian report:

. . . Late last year, the Medvezhe field was capable of producing 18 or 19 million cubic meters per day. They actually produced 10 to 11. The gas pipeline throughput capacity was the limiting factor.

. . . the first compressor or station was built in four months. They have been working on the second compressor station already eight months.

. . . The goals set last year for the construction of the second link of the gas pipeline in the southern sector and the expansion of compressor stations have not been completely fulfilled.[36]

The Ninth Five-Year Plan set the construction goals for gas pipelines at 30,000 kilometers. These pipelines would transport gas from new fields discovered in frozen regions of the Arctic and the hot deserts of Central Asia. The harsh climate and isolated locations have severely tested the Soviet gas industry, and performance from 1971 to 1973 fell short of assigned goals. According to Robert Ebel, new gas pipeline construction achieved only 85.5 percent of its goal in 1971 and 84 percent in 1972.[37]

Compounding the failure to meet the planned goals for pipeline construction was the problem of cost overruns. Construction costs estimated at the beginning of the Ninth Five-Year Plan have escalated from 25 to 100 percent over original cost calculations because of planning errors, increases in purchase price of equipment required, and changes necessitated by environmental conditions.[38]

The sheer magnitude of the problems in the Arctic has forced the Soviets to back away from permafrost whenever possible. In Arctic pipeline construction, the Soviets appear to be moving away from installing the pipe on supports. The Medvezhe-Punga line was laid on the surface or trenched. Also, the entire parallel line laid alongside the Messoyaka-Norilsk pipeline is being trenched. This line is the northernmost gas pipeline in the world, and it has been plagued by frequent failure of materials and equipment.[39]

Soviet North Star negotiators have suggested to their American counterparts that the location of the proposed North Star pipeline be moved south, just below the permafrost region. United States negotiators have refused to accept the Soviet suggestion because it involved large increases in cost. The Americans are convinced that the pipeline can be operated successfully in the permafrost region.[40]

Because of the harshness of the Arctic environment, construction work must be limited to the winter months when the ground is frozen. However, the extreme cold reduces worker efficiency. In May 1968, the chief of the Tyumen Petroleum and Gas Construction Administration attempted to maximize output during the short construction period by employing the concept of modular construction. A plant was established to build sections for Arctic installations. Equipment for boiler houses, gas-distributing stations, water stations, and transformer substations was prefabricated at the plant and delivered in sealed containers to the construction site for installation. Through this process, labor on site was reduced by 80 to 90 percent, with a reduction in costs of approximately 15 to 25 percent and a substantial improvement in the quality of work. This procedure was used for the first 580 kilometers of the Medvezhe-Nadym-Punga gas pipeline, which was completed in only four months. The gas collection station number two at the Medvezhe field was constructed in four months with a construction labor force of only nine men per shift. The official construction norms allowed 22 months for the project.[41]

The new construction procedures are not without problems. When modular units were delivered recently to a construction site in the north, there was no labor force to install the modules. The construction workers refused to do the job, because there was no approved procedure to pay them for the work. No material incentives existed for the new procedures, and construction workers found it more profitable to employ the old, antiquated, and more expensive procedures.[42]

Despite the great difficulty and expense of Arctic operations, the Medvezhe field was brought on-line with the completion of the Medvezhe-Punga pipeline in 1973. A second pipeline between Punga and Medvezhe is being laid in order to handle the anticipated increase in output from Medvezhe by 1975.[43]

In summary, Arctic operations pose almost insurmountable obstacles for the Soviet manager. Technology is needed in almost every aspect of operations including metallurgy, casings, drill bits, compressors, and pipeline. Despite the Soviet technology gap, gas output was increased by 15 BCM, or almost 8 percent, in 1973. The Soviets have demonstrated the ability to work around obstacles and to get the job done. There is no doubt that their task would be simplified, increased in efficiency and productivity, and reduced in cost if they had access to Western technology. They will accomplish their energy goals, however, with or without assistance from the United States.

SUMMARY

Soviet technology lags behind that available in the West. World gas technology is advancing very rapidly, and no single nation enjoys a monopoly. All nations would benefit, however, from a continuing world exchange of gas technology.

Perhaps the Soviet Union would benefit most from this exchange because the task of developing Arctic gas resources would be simplified, increased in efficiency and productivity, and reduced in cost if they had access to Western technology. The Soviets could contribute to world gas technology in that they have built and operated gas pipelines in the Arctic since 1964. Also, the Soviets possess the only operational wide-diameter (56-inch) pipeline in the world. It is in regular operation in Western Siberia.

The United States would be the most important supplier of technology in the sense of providing metallurgy, drill bits, casings, compressors, and related equipment. There would be benefits, however, in that the United States would be forced to develop a capability to manufacture wide-diameter pipe—an area of gas technology in which the United States is noncompetitive at this time.

NOTES

1. Central Intelligence Agency, Soviet Economy in 1973, p. 1.

2. Leon Herman, "The Promise of Economic Self-Sufficiency under Soviet Socialism," in The Development of the Soviet Economy (New York: Praeger Publishers, 1968), p. 216.

3. N. K. Baibakov, "O Gosudarstvennom Plane Razvitia Narodnovo Khozyajstva SSSR na 1974 God" [Concerning the Government Plan for Development of the National Economy of the USSR in 1974], Pravda, December 13, 1973, p. 3.

4. Michael Boretsky, "Comparative Progress in Technology, Productivity and Economic Efficiency: USSR versus USA," in U.S. Congress, Joint Economic Committee, New Directions in the Soviet Economy, pt. 2-A, Economic Performance, 89th Cong., 2nd sess., 1966, p. 149.

5. Gertude Schroeder, remarks made during the East-West Economic Developments Seminar of the Joint Symposium on "Soviet Power and Europe," sponsored by the Association for Advancement of Slavic Studies (AAASS), Washington Chapter, and Institute for Sino-Soviet Studies, The George Washington University, May 10, 1974.

6. U.S. Department of Commerce, U.S. Commercial Relationship in a New Era, prepared by Peter G. Peterson (Washington, D.C.: Government Printing Office, August 1972), pp. 57-59.

7. U.S. Congress, House, Committee on Foreign Affairs, U.S.-Soviet Commercial Relations: The Interplay of Economics, Technology Transfer, and Diplomacy, prepared for the Subcommittee on

National Security Policy and Scientific Developments by John P. Hardt
and George D. Holliday (Washington, D.C.: Government Printing
Office, 1973), pp. 15-16. *H382-23*

8. U.S. Department of Commerce, op. cit., p. 64.

9. Baibakov, op. cit., p. 3.

10. Robert W. Campbell, "Technology Transfer in Expanded
Commercial Relations between the U.S. and USSR," testimony in
House Subcommittee Hearings on International Economic Policy,
1974, pp. 183-88. *H291-156.*

11. J. Fred Bucy, testimony, in U.S. Congress, House, Committee
on Science and Astronautics, The Technology Balance: U.S.-USSR
Advanced Technology Transfer, Hearings before the Subcommittee on
International Cooperation in Science and Space, 93rd Cong., 1st and
2nd sess., December 1973, p. 53.

12. Anthony C. Sutton, testimony in House Subcommittee Hearings
on International Economic Policy, 1974, pp. 156-58.

13. John P. Hardt, "Summary," in U.S. Congress, Joint Economic
Committee, Soviet Economic Prospects for the Seventies (1973), pp.
xi-xii.

14. Campbell, op. cit.,

15. Jack H. Ray, interview held in Washington, D.C., November
1973.

16. Anthony C. Sutton, Western Technology and Soviet Economic
Development, 1917 to 1930 (Stanford, Calif.: Hoover Institution, 1968),
pp. 347-348.

17. Dan Walsh, interview and correspondence during May and
June 1974.

18. Dean Hale, "International LNG Movements Continue to
Expand," Pipeline and Gas Journal (December 1973): 61-63.

19. "Summary of Long-Term LNG Import Projects," Gas Supply
Review, December 15, 1973, p. 22.

20. The information in this paragraph was derived from interviews and correspondence with Jack H. Ray over the period September
1973 to August 1974. Also, the information was presented in hearings
before the Senate Subcommittee on Multinational Corporations, Committee on Foreign Relations, on June 17, 1974.

21. Dmitri V. Belorusov, et al., Osvoyeniye Neftyanykh
Mestorozhdenij Zapadnoj Sibiri [Mastering the Petroleum Fields in
Western Siberia] (Moscow: Nedra, 1972), p. 178.

22. John P. Hardt, "Western Siberia: The Quest for Energy,"
Problems of Communism (May-June 1973): 27-29.

23. Pravda, March 5, 1973, p. 2; and Izvestia, February 3, 1973,
p. 1; March 20, 1973, p. 1; and March 21, 1973, p. 3.

24. "Gazovaya Promyshlennost" [Gas Industry], Ekonomicheskaya
Gazeta [The Economic Gazette], no. 8 (February 1973): 2.

25. J. Richard Lee, "The Soviet Petroleum Industry: Promise and Problems," in U.S. Congress, Joint Economic Committee, Soviet Economic Prospects for the Seventies (1973), pp. 284-86.

26. Ibid., p. 289.

27. Petroleum Extension Service, Oil Pipeline Construction and Maintenance (Austin: University of Texas, April 1973), p. 3.

28. U.S. Natural Gas Delegation, Gas in the Soviet Union (Arlington, Va.: American Gas Association, Inc., 1970), pp. 26-29; and Robert W. Campbell, "Some Issues in Soviet Energy Policy for the Seventies," in U.S. Congress, Joint Economic Committee, Soviet Economic Prospects for the Seventies: A Compendium of Papers Submitted to the Joint Economic Committee, ed. John P. Hardt (Washington, D.C.: Government Printing Office, 1973), p. 48.

29. V. Sukhanov and E. Shatokhin, "Bolshoi Gaz Zapolyar'ya" [Huge Gas of the Arctic], Izvestia, March 23, 1973, p. 3.

30. Belorusov, et al., op. cit., pp. 126-129.

31. Ibid.

32. Ibid., pp. 134-36.

33. Ibid.

34. Robert W. Campbell, The Economics of Soviet Oil and Gas (Baltimore: Johns Hopkins Press, 1968), p. 152.

35. Lee, op. cit., p. 289; and Campbell, op. cit., p. 48.

36. V. Sukhanov and E. Shatokhin, "Tockka Otchyeta: Zapadno-Sibirokiz Kompleks: Opyt i Problemy [Bench Mark: West-Siberian Complex: Experience and Problems], Izvestia, March 20, 1973, p. 2.

37. Robert E. Ebel, "Russia Falls Short of Pipe Line Goals," Pipeline Industry, November 1973, pp. 33-36.

38. Ibid.

39. Ibid.

40. Interview with Ewell H. Muse, III, manager of economic planning, Tennessee Gas Transmission, held in the Washington, D.C., office of Tenneco Inc., April 21, 1974.

41. Sukhanov and Shatokhin, "Tochka Otchyeta," op. cit., p. 2.

42. Ibid.

43. Ebel, op. cit., pp. 33-36.

No equity will pass under the North Star proposal, because Soviet law precludes private ownership of the means of production. As "owners" the Soviets will bear the risk of ownership, and U.S. investments will constitute a loan or advance production payment to be repaid in the form of natural gas output. Should the project fail, the Soviet government would be obligated to repay the funds invested.

In view of the large investments needed to implement North Star and the risk of developing natural gas resources in the hostile Arctic environment, prudence indicates that the Soviet credit worthiness or ability to repay the investments be analyzed.

CONSIDERATIONS OF THE SOVIET BALANCE OF PAYMENTS

In a planned economy, such as exists in the Soviet Union, it is not possible to view the North Star project in isolation. Every aspect of the economy is interconnected. In this project, domestic and international factors are intimately related, and the relationship is reflected in the Soviet balance of payments.

The North Star project is affected by the balance of payments in two ways. First, the foreign creditor is concerned about the repayment of his loans. The extraction of natural gas from the hostile Siberian environment involves an element of technological risk. Should the project not succeed in delivering gas according to the terms of the yet-to-be-specified agreement, how would loans be repaid? The Soviet government could repay the debt either from Soviet gold reserves or from current earnings. The latter capability is a function of the Soviet debt service ratio, that is, the percentage of hard-currency earnings required to repay the total principal and interest due on all foreign borrowings.

The second effect of balance of payments on North Star is the potential for hard-currency earnings if the project performs as planned. These credits would permit Soviet purchases of Western goods needed by the domestic economy. During the decade of the 1960s, Soviet trade was characterized by persistent deficits in accounts with the West. Published Soviet plans indicate a desire to continue imports of Western equipment and materials. Potential hard-currency earnings from North Star could provide an incentive to the Soviets, both to sign the agreement and to meet its delivery terms after the agreement is signed.

Soviet policy, particularly over the past decade, has been directed at the expansion of trade with Western developed countries for the purpose of importing equipment, technology, and scarce materials. The USSR minister of foreign trade stated recently that rapid technical progress cannot be accomplished unless there is a broad utilization of world discoveries in various branches of production and unless there is a mutual exchange of these discoveries, that is, the development of extensive "international division of labor."[1] His remarks were as follows:

> Exchanges of machinery and equipment are very important for Soviet trade with the countries of Western Europe. For many years the USSR has been a major customer for West European machinery and equipment of the chemical and wood-processing industries, for motor vehicle plants, building materials industry, light industry, and food industry enterprises.
>
> Agreements on cooperation with firms of the West European countries for the development of the USSR's natural resources and the building on our territory of installations for the production of industrial products on a compensatory basis are a comparatively new form of foreign economic tie.
>
> The essence of these agreements is that foreign credits are provided for a designated purpose, credits that Soviet organizations use to acquire equipment, materials, and the necessary licenses, and the credits are eventually repaid from the production of the newly built facilities.[2]

The North Star project is one of these "comparatively new forms of economic tie." The project would provide equipment and assistance to develop natural gas resources and would receive payment in natural gas. An added advantage of the project is that output would be larger than that necessary to service the debt, and a surplus of some $8 billion in residual cash funds would be generated. All proceeds over the full life of the project would remain in the United States. After debt service, the residual funds would have to be spent for other U.S. goods and services.[3] The Soviet negotiators have not balked at this arrangement, perhaps because it would ease the Soviet hard-currency deficit, which averaged about $250 million annually during the decade of the 1960s and increased sharply to almost $1.8 billion in 1973.[4] Table 8.1 shows the trend in Soviet hard-currency deficits.

TABLE 8.1

Hard-Currency Trade Deficits of the Soviet Union
(in millions of U.S. dollars)

Year	Exports	Imports	Deficit
1960	768	1,018	−250
1961	900	1,061	−161
1962	951	1,184	−233
1963	1,012	1,287	−275
1964	1,073	1,156	−483
1965	1,374	1,569	−186
1966	1,516	1,755	−238
1967	1,711	1,616	+95
1968	1,909	2,018	−109
1969	2,125	2,436	−311
1970	2,197	2,711	−514
1971	2,652	2,955	−303
1972	2,815	4,171	−1,356
1973	4,817	6,566	−1,749

Sources: John T. Farrell, "Soviet Payments Problems in Trade with the West," in U.S. Congress, Joint Economic Committee, Soviet Economic Prospects for the Seventies, ed. John P. Hardt (Washington, D.C.: Government Printing Office, 1973), p. 702; Robert W. Kovach, "USSR Hard Currency Balance of Payments," presentation given at the U.S. Department of Agriculture, November 30, 1973; USSR, Vneshnyaha Torgovliya [Foreign Trade], Soviet yearbooks from 1960 through 1973; and interviews and correspondence with John T. Farrell and Robert S. Kovach between September 1973 and August 1974.

SOVIET HARD-CURRENCY DEFICITS

During the period 1960 to 1965, hard-currency trade deficits were financed primarily by the sale of gold. The gold transactions were conducted at an economic loss. Western experts estimate that the cost of producing one ounce of Soviet gold amounted to approximately $80, and the gold was sold at the official price of $35 an ounce.[5] Acceptance of this economic penalty reflects the importance that the Russians have placed on Western imports.

The Russians ceased to sell gold in 1965 when their reserves decreased to approximately 1,000 tons and when Western credit loans became available as an alternative means of financing imports. During the period 1966 through 1971, hard-currency trade deficits were financed chiefly by foreign credit loans. In 1972, the hard-currency deficit rose sharply to a record $1.4 billion as a result of the large imports of grain and agricultural products. Credit borrowings included approximately $500 million in Western government-backed loans to pay for imports of machinery, equipment, and pipe. In addition, approximately $100 million in Commodity Credit Corporation three-year-credit loans were borrowed to finance purchases of grain from the United States. As a result of these loans, the total outstanding Russian debt increased to an estimated amount of $2.6 billion.[6] The government-backed loans were not sufficient to cover the entire 1972 hard-currency deficit, and several hundred million dollars in medium- and short-term loans were borrowed from the Eurocurrency market. The balance of the deficit was covered by the sale of gold.[7] Debt service payments on the accumulated foreign loan obligations grew steadily from 1960, and by 1972 the debt service ratio reached 20 percent of the Soviet hard-currency export earnings. In 1973, debt service increased to $814 million, and the debt service ratio fell to 17 percent of hard-currency exports.[8] This improvement was due in large part to the world price inflation in commodities resulting in increased Soviet earnings on petroleum and other raw material exports. The performance trend of the Soviet foreign debt burden is indicated in Table 8.2

Part of the Soviet hard-currency trade deficit was financed by Western government-backed credit loans. This method of financing imports reduces Russian flexibility in importing desired Western goods because export earnings are mortgaged to pay for purchases made in previous years. A more important aspect of the increasing Soviet debt service ratio is that it could limit or cut off a foreign creditor's willingness to lend. The debt service level at which cutoff occurs is usually 25 to 30 percent of export earnings.[9]

TABLE 8.2

' Foreign Debt Burden of the USSR
(in millions of U.S. dollars)

Year	Hard-Currency Exports	Debt Service Principal and Interest	Debt Service Ratio (in percent)
1960	768	39	5
1961	900	76	8
1962	951	116	12
1963	1,012	144	14
1964	1,073	162	15
1965	1,374	166	12
1966	1,516	170	11
1967	1,711	181	11
1968	1,909	255	13
1969	2,125	322	15
1970	2,197	379	18
1971	2,652	463	18
1972	2 2,815	562	20
1973	4,817	814	17

Sources: Robert S. Kovach, "USSR Hard-Currency Balance of Payments," presentation given at U.S. Department of Agriculture, November 30, 1973; and John T. Farrell, "Soviet Payments Problems in Trade with the West," in U.S. Congress, Joint Economic Committee, Soviet Economic Prospects for the Seventies, ed. John P. Hardt (Washington, D.C.: Government Printing Office, 1973), p. 702; and informal discussions with both authors. Data were verified by spot checks in various issues of USSR, Ministry of Foreign Trade, Vneshnyaya Torgovlya [Foreign Trade of the USSR], the statistical yearbook on trade.

The excellent credit rating established by the Soviet debt-payment record since the end of World War II and the self-liquidating nature of part of the Soviet debt combined with the large Soviet gold reserves suggest that a 30 to 50 percent debt service ratio could be tolerated. Continued Soviet borrowing to cover trade deficits, however, could ultimately jeopardize the Russian preferred credit rating. The trend in Soviet drawings and repayments on loans guaranteed by Western governments is indicated in Table 8.3.

TABLE 8.3

Soviet Drawings and Repayments on Loans
Guaranteed by Western Governments
(in millions of U.S. dollars)

Year	Hard-Currency Deficits	New Credit Loans	Repayments		Total Foreign Debt
			Principal	Interest	
1960	-250	125	37	2	136
1961	-161	165	70	6	231
1962	-233	180	106	10	305
1963	-275	140	130	14	315
1964	-483	170	147	15	338
1965	-186	190	149	17	280
1966	-238	275	150	20	505
1967	+ 95	305	152	29	658
1968	-109	510	217	38	951
1969	-311	630	265	57	1,316
1970	-514	715	310	79	1,722
1971	-303	682	374	103	2,029
1972	-1,356	1,030	451	122	2,608
1973	-1,749	1,690	657	157	3,641

Sources: Robert S. Kovach, "USSR Hard-Currency Balance of Payments," presentation given at U.S. Department of Agriculture, November 30, 1973; and John T. Farrell, "Soviet Payments Problems in Trade with the West," in U.S. Congress, Joint Economic Committee, Soviet Economic Prospects for the Seventies, ed. John P. Hardt (Washington, D.C.: Government Printing Office, 1973), pp. 690-72; and informal discussions with both authors. Data were verified by spot checks in various issues of USSR, Ministry of Foreign Trade, Vneshnyaya Torgovlya [Foreign Trade of the USSR], the statistical yearbook on trade. Hard currency countries are those listed by the Bank of International Settlements as trading in hard currency—generally the Western developed countries, less Finland.

The large hard-currency trade deficit, the desire to preserve their excellent credit rating, the desire to avoid the high interest rates of the Eurocurrency market, and the sharp increase in the free-market price of gold all pointed to the sale of gold as the logical course of action for the solution of the Soviet hard-currency deficit problem. And this course of action was adopted by the Soviet leadership. It is estimated that the Soviets sold 150 tons of gold in 1972 at an average price of $60 an ounce[10] and 280 tons in 1973 at an average price of $90 an ounce.[11] Most of the 1973 gold sales occurred during the period of strong demand in the first three quarters. There was little evidence of Soviet sale of gold during the fourth quarter of 1973.[12] A deeper analysis of the Soviet capability to make gold transactions is needed to determine whether this course of action has application in the North Star project.

SOVIET GOLD RESERVES

The gold reserves and the annual production of gold in the Soviet Union are considered state secrets. However, the Five-Year Plan and discussions in the press indicate that efforts are being made to expand the production of gold. In 1971, annual production was estimated at 6.7 million troy ounces (210 short tons),* and the production was forecast to increase to 7.7 million troy ounces (246 short tons) in 1975, and to 8.9 million ounces (278 short tons) by 1980.[13]

If past history is used as a guide for Soviet gold transactions, the Soviets appear willing to sell gold to aid in the solution of critical domestic problems. They have allowed their gold reserves to fall to 1,000 tons. This action was taken when poor harvests, in the 1963-65 period, forced the Soviets to make large grain purchases in the West. As gold reserves dropped to 1,000 tons, Soviet policy turned to a reduction of imports as a solution to the hard-currency deficit problem. By 1967, this policy resulted in a $95-million surplus in the hard-currency trade accounts, as shown in Table 8.4.

During the period 1966 through 1971, trade with the West continued, with financing provided by increasing Soviet exports of manufactured goods and by loans granted or guaranteed by foreign governments. After 1971, as Soviet gold reserves were replenished, gold sales were resumed. The trends of Russian gold sales and gold reserves are likewise indicated in Table 8.4.

Since 1972, Soviet gold sales indicate a willingness to sell gold when the market price is high. There appears to be no hesitation to sell off-current production, or to dip into reserves, if the market

---

*Data in this study are computed on the basis of short tons of 2,000 pounds. Some references use gross tons of 2,240 pounds or metric tons of 2,205 pounds. The unit of measurement used accounts for some quantity discrepancy found in the literature.

TABLE 8.4

Relationship of Hard-Currency Deficits to
Soviet Gold Sales and Gold Reserves
(in millions of U.S. dollars)

Year	Hard-Currency Deficits	Gold Sales		Gold Reserves	
		Dollars	Tons	Dollars	Tons
1960	-250	200	180	2,555	2,270
1961	-161	300	270	2,365	2,100
1962	-233	215	195	2,250	2,000
1963	-275	550	500	1,800	1,600
1964	-483	450	410	1,495	1,330
1965	-186	550	500	1,095	975
1966	-238	Nil	Nil	1,265	1,125
1967	+95	15	14	1,425	1,265
1968	-109	12	11	1,590	1,425
1969	-311	Nil	Nil	1,765	1,570
1970	-514	Nil	Nil	1,945	1,730
1971	-303	Nil	Nil	2,135	1,895
1972	-1,356	275	150	3,744	1,950
1973	-1,749	824	280	5,832	1,800

Note: Gold prices are based on the official price of gold, $35 an ounce, through 1971. The 1972 conversion rate for sales and reserves is $60 an ounce, and for 1973 the conversion rate used was $90 an ounce. These prices were the estimated prices at which the Soviets actually sold gold in the respective years. The average 1973 price of gold was $97.21 an ounce, based on daily fixing in U.S. dollars, and the high price for the year was $127 an ounce (Bullion Review, p. 16). In view of the current high free market price of gold, the value cited for Soviet gold reserves in 1973 is conservative.

Sources: Robert S. Kovach, "USSR Hard-Currency Balance of Payments," presentation made at U.S. Department of Agriculture, November 30, 1973; and John T. Farrell, "Soviet Payments Problems in Trade with the West," in U.S. Congress, Joint Economic Committee, Soviet Economic Prospects for the Seventies, ed. John P. Hardt (Washington, D.C.: Government Printing Office, 1973), p. 702; and informal discussions with both authors. Data were verified by spot checks in various issues of USSR, Ministry of Foreign Trade, Vneshnyaya Torgovlya [Foreign Trade of the USSR], the statistical yearbook on trade. Data are supplemented with information from Samuel Montagu and Co. Ltd., Annual Bullion Review, 1973 (London, March 1974), and "Soviets Sold 280 Tons of Gold in '73," Washington Post, May 18, 1974, p. D-7.

price of gold is sufficiently high. On this basis, the sale of current gold production, at an estimated market price of $100 an ounce and assuming 80 tons for domestic consumption, would net approximately $900 million annually, by 1980.

Two additional trends are exerting a favorable influence on the Soviet hard-currency trade balance: (1) commodity price trends and (2) arms sales.

Soviet imports from the West consist of about two-thirds machinery and technology and one-third agricultural products (the percentages were reversed in 1972-73 because of the poor Soviet harvest). Approximately three-fourths of Soviet exports consist of fuels and raw materials, such as oil, gas, ores, and metals.[14] (See Appendix for commodities exchanged in U.S.-Soviet trade.) The trend of the world price in primary products has been moving sharply upward, while the price of manufactures has been increasing slowly. These trends, if they continue, will accrue to the Soviet benefit by reducing the hard-currency trade deficits through price adjustments.

The second factor that has an indirect influence on the Soviet hard-currency trade balance is the sale of arms to the Arab nations during the 1973 Arab-Israeli war and the future augmentation of these arms supplies. Arab revenues from the increase in oil prices will permit the payment for arms to be made in hard currencies. The arms sales are classified, and accurate data on quantities and prices are not available. However, these developments do exert a strongly positive influence in solving the hard-currency deficit problem.

The Soviet Union introduced the concept of using Western credits to build production facilities and develop natural resource endowments a number of years ago. This arrangement aided in solving the hard-currency deficit problem. By tying credit repayments to the productive output of newly created facilities, the loans became self-liquidating. A series of such agreements has been concluded, involving a total of $1.3 billion in credits—$1,005 million in pipe and pipeline equipment and $260 million in Siberian timber and port development projects.[15] Agreements have been signed with Austria, West Germany, Italy, and France to supply pipe and equipment on credit varying from 5 to 10 years in length, with repayment to be made in deliveries of natural gas under separate but related agreements. Similarly, Japanese credits for the port development at Vrangel Bay will be repaid in Soviet wood and wood chip deliveries.[16] Some $800 million of the total outstanding foreign debt of $3.6 billion[17] at the end of 1973 was tied to self-liquidating loans. By the time these projects are amortized, in 1984, they will have earned a $700-million credit surplus over and above loan repayments.[18]

SUMMARY

In this chapter, Soviet credit worthiness was analyzed in order to ascertain Soviet ability to repay the North Star investment should the project fail. The findings were as follows:

1. The Soviet Union has averaged a hard-currency deficit in trade with the West since 1960, and trade deficits have increased sharply since 1970.
2. Trade deficits have been financed by loans from Western governments or by the sale of gold. Gold reserves stood at 1,800 tons in 1973, or approximately $6 to $12 billion, depending on the price used for conversion.
3. World commodity inflation has helped to reduce the Soviet hard-currency trade deficit. Soviet debt service ratio increased from 5 percent in 1960 to 20 percent in 1972, then reduced to 17 percent in 1973 due to favorable commodity price trends.
4. Increases in the world price of gold and profits from the sale of arms to the Arabs have improved the Soviet hard-currency posture.
5. Adequate financial resources exist to permit repayment of North Star loans, should the project fail; however, Soviet trade with the West would probably be curtailed if the project were to fail.
6. If North Star is successfully implemented, surplus hard-currency earnings from the project would encourage expanded U.S.-Soviet trade.

NOTES

1. N. Patolichev, "Vzaimovygodnoye Sotrudnichestvo" [Mutually Advantageous Cooperation], Pravda, December 27, 1973, p. 4.
2. Ibid.
3. Testimony of Jack Ray at hearings before the Subcommittee on International Trade, in House Subcommittee Hearings on International Economic Policy, 1974, p. 216.
4. Robert S. Kovach, "USSR Hard-Currency Balance of Payments," presentation given at the U.S. Department of Agriculture, November 30, 1973.
5. Michael Kaser, "Soviet Union," International Currency Review (July-August 1973): 92.
6. Kovach, loc. cit.
7. Ibid.; and John T. Farrell, "Soviet Payments Problems in Trade with the West," in U.S. Congress, Joint Economic Committee,

Soviet Economic Prospects for the Seventies (1973), pp. 693-94,
702.

8. Interview with John Farrell in Washington, D.C., July 25, 1974;
and Douglas Whitehouse, presentation made at the Joint Symposium of
AAASS at The George Washington University, May 11, 1974.

9. "How's Brezhnev's Credit Rating?" Forbes, June 15, 1973, pp.
33-34.

10. Kovach, loc. cit.; and Farrell, op. cit., p. 702.

11. Samuel Montagu and Co. Ltd., Annual Bullion Review, 1973
(London, March 1974), p. 9; and "Soviets Sold 280 Tons of Gold in
'73," Washington Post, March 18, 1974, p. D-7.

12. Samuel Montagu and Co., ibid.

13. V. V. Strishkov, "The Minerals Industry of the USSR," preprint from the 1971 Bureau of Mines Minerals Yearbook, U.S. Department of the Interior (Washington, D.C.: Government Printing Office,
1971), p. 20.

14. USSR Ministry of Foreign Trade, Vneshnyaya Torgovlya SSSR
za 1972 God [Foreign Trade of the USSR for 1972] (Moscow:
Mezhdunarodnye Otnoshenia, 1973), p. 19.

15. Farrell, op. cit., pp. 693-94.

16. Raymond J. Albright, Siberian Energy for Japan and the
United States, case study for the Senior Seminar in Foreign Policy,
Department of State, 1972-73, p. 8.

17. John T. Farrell, interview, July 24, 1974.

18. Kovach, loc. cit.

III

INSTITUTIONAL
OBSTACLES TO THE
NORTH STAR PROJECT
WITHIN THE
UNITED STATES

In the United States, legal and organizational structures are designed to protect the separation of powers concept, but they also restrict certain desired actions that might affect the efficiency, or may even preclude consummation of the North Star project. The detente strategy undertaken by President Richard Nixon and continued by President Gerald Ford for improving relations with the Soviet Union, that is, the strategy of political-economic linkages, is very difficult to implement under this institutional framework. Certain policies involving trade and finance are not within the presidential power to establish. This authority is vested in the legislative branch of the government by the Constitution.

. . . Congress shall have the power to regulate commerce with foreign nations. (Article 1, Sec. 8)
The executive power shall be vested in a president.
. . . He shall have power, by and with the advice and consent of the Senate, to make treaties. (Article 2, Secs. 1 and 2)

Congress establishes the legal basis for foreign commercial relations, while the president executes and administers the legislation enacted. Close coordination between two large, complex, and independent branches of government is required. This separation of powers makes it very difficult to embark on major new initiatives in trade relations.

The established institutional arrangement places the United States at a disadvantage in dealing with the Soviet Union. For example, Soviet and U.S. negotiators worked out a trade agreement on October 18, 1972. Although parts of the agreement were implemented and trade has expanded dramatically, the legal base in the form of approved legislation was not enacted by the Congress until January 1975. On June 30, 1974, the Export Administration Act of 1969, which established the legal basis for conducting U.S. foreign trade relations, and the Export-Import Bank Act of 1945, which provides the financial base to support foreign trade, both expired. Both acts were extended with 30-day continuing resolutions, but the permanent legal basis for the conduct of U.S. foreign commercial relations was not signed by the president until January 4, 1975. Furthermore, congressional restrictions enacted in the legislation pose obstacles to the Trade Agreement of 1972, particularly limitations on extension of credits to the Soviet Union and clauses that would prohibit the

123

granting of most-favored-nation (MFN) treatment to any nonmarket economy country which denies its citizens the right to emigrate or which imposes more than nominal fees upon its citizens as a condition of emigration. Because of unacceptable conditions imposed by the congressional legislations, the Soviet Union announced that it would not implement the Trade Agreement of 1972.

Because the North Star project must operate within the legal, institutional, and ideological framework of the United States, an analysis will be made of the trade authorization and trade financing factors that constitute obstacles to implementation of this project.

9

There are three broad categories of trade authorization factors that bear on the North Star project: (1) the most-favored-nation (MFN) treatment, (2) the problem of dumping, and (3) the transfer of technology. They will be discussed separately to provide an understanding of their importance in the consideration of the project.

MOST-FAVORED-NATION (MFN) TREATMENT

The United States adopted a policy of unconditional MFN treatment in 1923. Under an MFN policy, a nation agrees to extend, automatically, to other nations any trade concession or advantage it grants to a third nation. After the Trade Agreements Act of 1934, the United States extended MFN routinely to all countries, whether or not a specific MFN agreement existed, provided other countries did not discriminate against U.S. products. In 1947, the United States signed the General Agreement on Tariffs and Trade (GATT) and assumed a contractual obligation to grant MFN treatment on a reciprocal basis to all other members of GATT. In 1951, specific legislation precluded extension of MFN to Communist nations.[1]

After the Great Depression of 1929-33, and particularly after World War II, U.S. trade policy moved world trade practices in the direction of multilateral free trade, but again excluded commercial relations with the Communist planned economies.[2]

The commercial procedures worked out by the free-market economies to promote world trade rely heavily on the market mechanism, whereby price acts as the "invisible hand" to regulate supply and demand. Trade theory accepts the premise that each nation enjoys a comparative advantage in producing certain products.

In order to increase trade on the basis of reciprocal advantage, the free-market nations have generally accepted a set of "rules of the game" to promote expanded trade, embodied in the GATT, which went into effect in 1948.[3] The broad concept embodied in these rules is that if each nation employs its resources in the area of its competitive advantage, output will be maximized and redistributed to participants through trade, to the mutual benefit of all. Imperfections in the theoretical model exist, as indicated by the exceptions permitted to Customs Unions, corrective actions allowed when a nation is in balance-of-payments difficulties, and the protection of domestic agriculture. Furthermore, national governments control the functioning of their respective domestic economies by regulating supply and demand through artificial barriers and stimuli. In particular, domestic economies are insulated from undesirable external influences by means of tariffs, quotas, and nontariff barriers. The success and benefits of the basic U.S. strategy of free trade, however, have resulted in an unprecedented expansion of world trade and the longest period of sustained and rapid income growth in history.[4] An important ingredient of this success story has been the willingness of the free-market nations to permit the adjustments in production and consumption necessary to achieve the benefits of comparative advantage.

The Soviet economy is a planned economy, which has not permitted the kinds of adjustments made in the free market. The plan dictates the details of production, consumption, allocation of resources, prices, and wages. There is no market mechanism whereby production and consumption can seek a balance through price. Prices are decreed by state planning organs. Furthermore, the plan specifies exactly what products are to be produced and by whom. Any change or failure in the plan causes major disruption, as the effect of the change ripples through the economy. The Soviets benefit from trade in that failures or errors in the plan can be corrected by purchase of short commodities in Western markets. It is not possible, however, for the Soviets to reciprocate by allowing Western purchasers to shop in their domestic market. Unplanned foreign purchases would disrupt the plan and cause major upheavals in the economy. For this reason, the Soviets do not allow foreigners to own Russian currency. The currency itself is not convertible. Most foreign trade is conducted through trade agreements on a bilateral basis. This arrangement forces trade to conform to the plan as closely as possible.[5] From the Soviet point of view, the North Star project is desirable. Its long-term predictable elements could be fit into the economic plan without serious disruption.

Most-Favored-Nation (MFN) Treatment
in Russian-U.S. Trade

The United States has conducted profitable trade with Russia from
the earliest days of the republic. Before 1790, U.S. ships made at
least 19 trips to Russia. Trade grew rapidly, and by 1802 there were
81 American ships engaged in highly profitable trade.[6] In 1811,
Russia purchased one-tenth of the U.S. exports, and the United States,
in turn, was heavily dependent on imports of Russian naval stores.[7]
Political factors, however, cast a shadow of uncertainty over the
commercial relations from this earliest period.

In 1780, Congress dispatched Francis Dana to Russia as a minister
to gain recognition for the republic and to negotiate a treaty of friendship and commerce. Dana was refused a reception by Catherine II,
and he departed St. Petersburg in August 1783, after two years of
the "most mortifying isolation."[8]

President George Washington nominated a consul to St. Petersburg
in 1794, and President John Adams again attempted to negotiate a
treaty of friendship and commerce in 1799. Official diplomatic relations were opened in 1809, when John Quincy Adams was confirmed
by the Senate as minister to Russia, and the Russians appointed Andre
Dashkoff as consul general at Philadelphia. Adams was unsuccessful,
however, in negotiating a trade treaty. Russia was deeply involved in
European politics during this period and was eventually forced into
war with Napoleon. It did not suit the Russian interest to enter into
a trade agreement with the United States at that time.[9]

A Treaty of Navigation and Commerce was finally signed on
December 18, 1832, and ratifications were exchanged in Washington
on May 11, 1833. The treaty provided for the mutual exchange of
MFN and contained the customary clauses covering personnel movements, shipping, and settlement of disputes.[10] The treaty provided
for cordial relations, which remained uninterrupted for a period of
80 years. The United States abrogated the treaty, effective January
1913, charging that the Russians had violated Article 1, which reads
as follows:

. . . The inhabitants of their respective States shall, mutually,
have liberty to enter the ports, places, and rivers of the
territories of each party, wherever foreign commerce is
permitted. They shall be at liberty to sojourn and reside in
all parts whatsoever of said territories, in order to attend
to their affairs, and they shall enjoy, to that effect, the same
security and protection as natives of the country wherein
they reside.[11]

The situation which led to the abrogation of the treaty by the United States involved the Russian treatment of Jewish emigrees. In the latter part of the nineteenth century, a large number of Jews emigrated from Russia, and a number of them came to the United States. Russian passport regulations restricted the entry and movement of foreign Jews in Russia. Some of the American emigrees, who were now U.S. citizens, desired to revisit Russia, and entry was denied. At the same time, Russian passport regulations excluded non-Talmudic Jews and Jews from the Middle Asiatic countries from the restrictions. A furor arose in the American press and led to the passage of a joint resolution in Congress which condemned the "refusal to honor American passports duly issued to American citizens, on account of race and religion." President Taft had little choice. He informed the Russians and abrogated the treaty, effective in January 1913.[12]

After the abrogation of the 1832 trade treaty, the United States continued to trade with Russia without the benefit of MFN. The Russian share of total U.S. imports dropped from 1.2 percent to less than 0.5 percent until the 1930s.[13]

In the 1930s, Stalin was deeply involved in a plan to modernize the Soviet economy and embarked on a program to import Western technology. At the same time, the Great Depression brought stagnation to industrial activity within the U.S. economy. The environment was ripe for negotiating a new trade agreement. The new commercial agreements were concluded on July 13, 1935. A State Department press release announced the granting of MFN with the following caveat:

. . . This agreement with the Soviet Union, although intimately related to the trade agreements program of the United States, was not concluded pursuant to the authority of the Trade Agreements Act of June 12, 1934. It does not involve any reciprocal concessions in respect to tariff rates. In return for the undertaking on the part of the Soviet government, which controls the import and export trade of the Soviet Union, to increase substantially its purchases of American products during the next twelve months, the government of the United States has agreed to extend to the Soviet Union, as long as the agreement remains in force, the benefits of tariff concessions granted under reciprocal trade agreements with other countries.[14]

The text of the agreement highlights the problem of MFN in trade agreements between market and planned economies, and particularly the problem of Soviet monopoly control of foreign trade. Trade in a free-market economy implies lower cost and increased consumption, because foreign competition has forced an adjustment in domestic prices and, perhaps, a shift in domestic production. In a planned economy, prices are controlled and not permitted to change. The effect of granting MFN to Russia would be to permit free entry of Soviet goods into the American market, but no reciprocal advantage would accrue to American producers, unless the "plan" were adjusted. Therefore, the trade agreement specified what adjustment was to be made to the plan; that is, by what volume Soviet imports would increase. The Soviet Union agreed to increase imports of American goods to a level of $30 million during the following 12 months. The figure represented an increase of over 100 percent over 1934 imports. In return, the U.S. government agreed to extend to the Soviet Union the tariff concessions which had been granted to Belgium, Haiti, and Sweden. On August 6, 1937, a new commercial agreement was announced, which granted the Soviet Union "unconditional and unrestricted most-favored-nation treatment" in return for a promise to purchase at least $40 million in American goods during the following 12 months.[15] Annual extensions of this agreement covered U.S.-Soviet commercial relations until the signing of the Lend-Lease agreement in 1942, when it was extended indefinitely.[16]

With the advent of the cold war, controls on exports to Communist countries were imposed by the Export Control Act of 1949. In 1951, the Trade Agreements Extension Act revoked MFN status for Communist states,[17] and Section 231 of the Trade Expansion Act of 1962 required congressional authorization for the extension of MFN treatment to any Communist nation.

Although there has been a thaw in the cold war, and U.S. trade restrictions on exports to Communist nations have been gradually eased since 1956, MFN status has not been granted to the Soviet Union. The trade agreement negotiated between U.S. and Soviet representatives on October 18, 1972, provided for the extension of MFN to the Soviet Union.[18] President Nixon requested the enactment of implementing legislation, but his request was modified with conditional restrictions and presented to President Ford for signature on January 3, 1975.

A question arises about what would be the increase in Soviet exports to the United States if MFN were granted. A recent Tariff Commission study estimated the increase at approximately 11.5 percent.[19] Tariff rates and discrimination against Soviet exports tend to be low, because they consist of predominantly raw and semimanufactured materials. The share of U.S. imports from the USSR that were subject to substantial discrimination (more than 5 percent above MFN rates), however, increased from 10 percent in 1970 to 25 percent in 1972.[20] The increased discrimination was due to two factors. The first was the

implementation of the final two stages of the Kennedy round of tariff
reductions, which resulted in a greater spread between MFN and non-
MFN tariff rates. The second was an increasing volume of manufactured goods being imported from the USSR.[21]

The study pointed out that the granting of MFN by other Western
nations was usually accompanied by a bilateral trade agreement. It
was difficult to separate out what part of the resulting increase in
trade was due to MFN. The bilateral trade agreement itself was
found to be a significant variable in the computations conducted and
was estimated to account for between 2.8 and 11.5 percent of the total
increase in trade.[22]

The increase in trade resulting from extending MFN treatment to
the Soviet Union is expected to be minor in the quantitative sense.
Despite existing MFN discrimination, there has been a phenomenal
increase in U.S.-Soviet trade since 1970. Apparently, the Soviet
economy was in need of Western goods, and the leadership was willing
to pay the price demanded. It must be added that the easing of political
tensions, reflected in the availability of Western credits and the
proposed trade agreement of October 18, 1972, also influenced the
trade increase.

Nevertheless, the Soviets consider the failure of the United States
to grant MFN status an unfriendly act and a conscious decision to
discriminate against Soviet goods. A leading Soviet spokesman
referred to it as the major obstacle to improved long-range commercial relations between the two countries:

> Despite progress, the main legal and trade problem existing
> between the Soviet Union and the United States remains that
> of most-favored-nation standing. Further development of
> mutual trade will depend on whether the Soviet Union is granted
> most-favored-nation status. . . .
>   . . . the USSR must have equal possibilities with others
> in the U.S. market, without any discrimination, and conditions
> must be created for bringing the trade agreement into force.[23]

The granting of MFN to the Soviet Union has been viewed with
ideological fervor in the United States. It has been linked to the
correction of Soviet mistreatment of its citizens of Jewish descent.
This idealistic stand taken in support of human rights is one on
which the United States has refused to yield in the past. In 1913,
President Taft abrogated a trade agreement and withdrew MFN from
Russia because Russian passport regulations discriminated against
American citizens of Jewish ancestry. The United States refused to
sign a new trade agreement, in spite of the fact that Russia was an

ally during World War I, and was willing to accept any amount of economic penalty but would not yield on principle. Apparently, the strong commitment to idealism still prevails in the American public.

For a time, there appeared to be a solution to the ideological impasse. A compromise agreement was worked out, which was reflected in an exchange of letters between Secretary of State Henry A. Kissinger and Senator Henry M. Jackson on October 18, 1974. The agreement provided that the president would convey assurances to the Congress that the level of emigration from the Soviet Union would rise and that punitive measures against Soviet citizens requesting exit visas would cease. In return, Congress would amend the proposed legislation to enable the president to waive the MFN and credit restrictions for 18 months, with subsequent 12-month waivers subject to congressional approval. This agreement, apparently, was accepted by Secretary Leonid I. Brezhnev, Foreign Minister Andrei A. Gromyko, and Ambassador Anatoly F. Dobrynin.[24]

The Trade Act of 1974 was amended to include the terms of the compromise agreement; however, credits available to the Soviet Union were limited to $300 million over a four-year period. Additional restrictions were contained in the amendments to the Export-Import Bank Act of 1974, which required that a separate presidential determination be made that a project was in the U.S. national interest if the loan exceeded $50 million. Furthermore, congressional approval was required for any loan that exceeded $60 million and for any energy project loans in the Soviet Union that exceeded $25 million. The amendments effectively precluded the use of credits as a lever to promote expanded U.S.-Soviet trade.

The Trade Act of 1974 was signed into law by President Ford on January 3, 1975, and the Export-Import Bank Amendments of 1974 were signed on January 4, 1975.

Secretary Kissinger informed the Congress on January 14, 1975, that the Soviet Union had decided not to bring into force the Trade Agreement of 1972. The restrictions on MFN and credits were considered to be discriminatory and an unacceptable interference in Soviet internal affairs.

The granting of MFN to the Soviet Union has taken on a political rather than economic meaning. Frederick B. Dent, secretary of commerce, has called the issue a "psychological deterrent" to commercial relations.[25]

Relation of MFN to North Star

Tariffs pose no economic barrier to implementation of the North Star proposal because the import of liquefied natural gas (LNG) is listed as a "free" item in U.S. Tariff Schedules.[26] Denial of MFN to the Soviet Union, however, would have adverse psychological implications for U.S.-Soviet commercial relations. The North Star project involves large investments, and it is planned that duties and obligations of both parties will be specified in firm, legal contracts. The efficient operation of the project, however, requires willing and not forced cooperation. Failure to grant MFN would hinder the reaching of a favorable decision on North Star by the Soviet Union, but would not preclude Soviet approval of the project.

THE PROBLEM OF DUMPING

Dumping is a practice in which a nation sells a product abroad at a price that is lower than the cost of production or lower than the price of the product in domestic markets. The United States and most Western nations have legislation that imposes countervailing tariffs and penalties when dumping is uncovered.

The Soviet price structure makes dumping very difficult to handle. Domestic prices are decreed and do not necessarily reflect the cost of production. In international trade, the Soviet Union is a small participant in a large market and trades at world market prices. World market prices are not related at all to Soviet domestic prices. For example, before the devaluation of the ruble in 1961, Soviet export prices were so overvalued that all exports appeared to be sold at a loss. The export prices were only 10 percent of domestic prices when converted at the official exchange rate.[27] Furthermore, the exchange rate of the ruble is set artificially. The overvaluation or undervaluation of the ruble complicates comparisons between domestic and foreign trade prices.

Superimposed on the price structure is the Soviet monopoly trade system, whereby the export and import transactions are handled by the same organizational structure (Ministry of Foreign Trade). Export and import may not be independent transactions. In other words, a loss on an export transaction may be more than made up by the profit on an import transaction. For example, if a product is exported at one-half of the domestic price and the proceeds are used to purchase an imported product at one-third of the domestic price, the transactions net a profit for the Soviet economy.[28]

Despite the difficulties associated with its price and monopoly trade system, the Soviet Union has been charged with dumping, and investigations have been conducted by the U.S. Tariff Commission. In 1967, the Soviet Union was found guilty of dumping, and imports of pig iron became liable to substantial duties.[29] In 1968, titanium sponge from the USSR became liable to dumping duties.[30]

The United States is confronted with the problem of domestic market disruptions due to trade with Western as well as Eastern nations. There have been profound changes in the level and pattern of world production and world trade since World War II. United States policy remains committed to expanded trade and an open and equitable world economic system, but the goals of removal of tariff and non-tariff barriers include a provision for

> . . . introducing an improved international "safeguard" system relating to measures taken to avoid disruption of national markets through sudden increases in competition from imports.[31]

A market disruption provision is included in the proposed trade agreement negotiated by American and Soviet representatives on October 18, 1972. The pertinent section reads as follows:

> Through state trading monopolies, the Soviet Union controls both the importation and exportation of all goods. In the Soviet economy, costs and prices do not necessarily play the same role as they do in a market economy. Accordingly, the Soviets have agreed to a procedure under which, after consultations, they will not ship products to the United States which the United States government has advised will "cause, threaten, or contribute to disruption of its domestic market."[32]

Existing legislation, as well as the proposed trade agreement with the Soviet Union, appears to provide adequate safeguards to protect the American market from the dangers of dumping. Dumping occurs when a nation attempts to rid itself of excess production, usually in a "one-shot" or short-term deal. With specific regard to the North Star project, the proposed delivery of 2.1 billion cubic feet (BCF) of natural gas over a period of 25 years is an arrangement which, by its very nature, is unlikely to cause market disruptions. Furthermore, import of gas under the North Star proposal requires a prior decision regarding price from the Federal Power Commission (FPC). This decision is an additional safeguard against dumping. Therefore, dumping is not deemed to be a problem under the North Star proposal.

TRANSFER OF TECHNOLOGY

The U.S. policy regarding technology transfer is set forth in the Export Administration Act of 1969, as amended by the Equal Export Opportunity Act of 1972,[33] and the Trade Act of 1974.[34] On the basis of the provisions contained in the legislation, the obstacles to technology transfer needed to support North Star are expected to be primarily of a legal and administrative nature.

Policy sections of the current legislation that apply to technology transfer are

DECLARATION OF POLICY

SEC. 3. The Congress makes the following declarations:

(1) It is the policy of the United States both (A) to encourage trade with all countries with which we have diplomatic or trading relations, except those countries with which such trade has been determined by the President to be against the national interest, and (B) to restrict the export of goods and technology which would make a significant contribution to the military potential of any other nation or nations which would prove detrimental to the national security of the United States.

(2) It is the policy of the United States to use export controls (A) to the extent necessary to protect the domestic economy from the excessive drain of scarce materials and to reduce the serious inflationary impact of abnormal foreign demand, (B) to the extent necessary to further significantly the foreign policy of the United States and to fulfill its international responsibilities, and (C) to the extent necessary to exercise the necessary vigilance over exports from the standpoint of their significance to the national security of the United States.

(3) It is the policy of the United States (A) to formulate, reformulate, and apply any necessary controls to the maximum extent possible in cooperation with all nations with which the United States has defense treaty commitments, and (B) to formulate a unified trade control policy to be observed by all such nations.

(4) It is the policy of the United States to use its economic resources and trade potential to further the sound growth and stability of its economy as well as to further its national security and foreign policy objectives.

(5) It is the policy of the United States (A) to oppose restrictive trade practices or boycotts fostered or imposed by foreign countries against other countries friendly to the United States, and (B) to encourage and request domestic concerns engaged in the export of articles, materials, supplies, or information to refuse to take any action, including the furnishing of information or the signing of agreements, which has the effect of furthering or supporting the restrictive trade practices or boycotts fostered or imposed by any foreign country against another country friendly to the United States.

(6) It is the policy of the United States that the desirability of subjecting, or continuing to subject, particular articles, materials, or supplies, including technical data or other information, to United States export controls should be subjected to review by and consultation with representatives of appropriate United States government agencies and qualified experts from private industry.[35]

The Trade Act of 1974 established an East-West Trade Board with the assigned task of monitoring trade between the United States and nonmarket economies to ensure that it is in the national interest of the United States. Specific provisions of the act provide that

SEC. 411. EAST-WEST FOREIGN TRADE BOARD
. . . (b)(1) Any person who exports technology vital to the national interest of the United States to a nonmarket economy country or an instrumentality of such country, and any agency of the United States which provides credits, guarantees, or insurance to such country or such instrumentality in an amount in excess of $5,000,000 during any calendar year, shall file a report with the board in such form and manner as the board requires which describes the nature and terms of such export or such provision.

(2) For purposes of paragraph (1), if the total amount of credits, guarantees, and insurance which an agency of the United States provides to all nonmarket economy countries and the instrumentalities of such countries exceeds $5,000,000 during a calendar year, then all subsequent provisions of credits, guarantees, or insurance in any amount, during such year shall be reported to the board under the provisions of paragraph (1).

(c) The board shall submit to Congress a quarterly report on trade between the United States and nonmarket economy countries and instrumentalities of such countries. Such

report shall include a review of the status of negotiations of
bilateral trade agreements between the United States and
such countries under this title, the activities of joint trade
commissions created pursuant to such agreements, the resolution of commercial disputes between the United States and
such countries, any exports from such countries which have
caused disruption of United States markets, and recommenda-
tions for the promotion of East-West trade in the national
interest of the United States.[36]

Although U.S. law supports expanded trade, including the export of
technology, it opposes the export of that technology which would
improve the military potential of any nation that might prove to be
damaging to the national security of the United States. The law provides for interagency review procedures to guard against this danger.

The Office of Export Administration in the Department of Commerce has primary responsibility for administering export controls.
Regulations require that exports of goods or technology to any
country except Canada be licensed. Usually, a general license is
issued and automatically provides general authorization for the export
of goods and technology covered by the license. Sensitive goods and
technology require a validated license granting specific authorization
for the type, quantity, and destination of goods or technology to be
exported.

Requests for license to export sensitive goods and technology are
subject to review by the Interagency Review Board before the license
is approved. Individual government agencies exercise primary jurisdiction over exports within their jurisdiction, such as Department of
Defense (military), Department of State (foreign policy), FPC (gas
and electric energy), and Energy Research and Development Administration (nuclear).

Individual agency interests are protected by separate legislation.
For example, the role of the Department of Defense in defending its
interests before the Interagency Review Board was strengthened by
sections included in the Defense Appropriations Authorization Act of
1975 (passed August 5, 1974) and the Export Administration Amendments (passed October 29, 1974). Prior to the passage of these acts,
agency conflicts over issue of an export license were resolved by
the president, and his decision was final. Under present law, the
secretary of defense must review all requests for export of technol-
ogies developed with Department of Defense funds. If he disagrees
with the issue of an export license and is overruled by the president,
that presidential decision may be overruled by a concurrent resolution
of Congress. In addition, the secretary of defense is authorized to

review a request for export license that might "significantly increase the potential military capability of a controlled country"; however, if the secretary of defense disagrees with the issue of an export license and is overruled by the president, the presidential decision is final.

Certain aspects of technology to be transferred under North Star would be subject to interagency review before export approval is granted. The question must be resolved whether the technology transferred contributes to the military capability of the USSR. The answer to that question is beyond the scope of this study, as it involves highly technical and political considerations. Data presented in Chapter 7, however, demonstrate that the technology required for North Star is available in Japan and Western Europe and that the Soviet Union has already demonstrated the ability to extract gas from Siberia. Therefore, the decision becomes a matter of subjective judgment rather than one of substance.

To add perspective to the problem of technology transfer to the USSR, it is important to be aware of the size of the U.S. contribution to Soviet domestic investment. In 1973, the Soviet gross national product (GNP) was approximately $600 billion. The Soviets reinvested one-third of this product, or approximately $200 billion, in the economy.[37] Total U.S. exports to the USSR that year were approximately $1.2 billion, of which $842 million consisted of food and grain.[38] The American contribution to Soviet domestic investment resources constituted between 0.2 and 0.5 percent of total investment resources available in 1973. The returns on technology spending can be expected to yield greater returns than investment in capital replacement. The size of American resources made available to Soviet managers, however, can make only a limited contribution to Soviet development plans.

Recent testimony before congressional investigating committees reveals that the Soviet military establishment lags behind the United States in advanced technology:

> We currently have a clear margin of advantage over the Soviets in this regard. For example, the USSR's technology applicable to nuclear weapons delivery systems lags significantly behind that of the West—from two to six years for such items as liquid and solid propulsion systems, guidance systems, integrated circuits, and advanced computers.[39]

Advanced technology was recognized as an asset which must be carefully managed; however, Roger E. Shields, a Department of Defense spokesman, agreed that it should not be embargoed absolutely:

Advanced technology, in short, is a major U.S. military,
economic, and diplomatic asset. As an asset, it should be
carefully managed. This does not imply that it should be
indiscriminately hoarded, but it does suggest that when we
export technology, we should assure that a clear and com-
mensurate return to the United States will be obtained.[40]

The impact of increased technological trade with the Soviet Union
on U.S. national security is not taken lightly by Americans. Much of
today's technology is multipurpose and has potential for military as
well as civilian applications. The specific responsibility for determining the danger posed to national security by technology transfer
is assigned to the Commerce Department. Requests for the transfer
of gas technology to the Soviet Union under the North Star project
will inevitably fall under the interagency review process. Although
such requests will most likely be approved, the review process can
be expected to cause time delays which will affect the cost and efficiency of implementing the North Star project.

U.S. Status in World Technology Transfer

For many years, the United States has led the world in research
activity and in the application of technology for commercial purposes.
In 1973, payments to the United States for patent royalties and management fees amounted to $5.5 billion, while U.S. payments to foreigners
for the same purpose amounted to $0.3 billion.[41]
Despite its world leadership role, the United States does not exercise a monopoly over world technology. Many creative developments,
which have revolutionized industrial processes, originated overseas.
They were imported and further developed by U.S. firms. Examples
include oxygen process steel, jet engines, and continuous casting to
float glass. The trend of foreign capability in technology is reflected
in the data covering the share of U.S. patents issued to four foreign
countries: West Germany, the United Kingdom, France, and Japan.
United States patents issued to these countries increased from 12 percent in 1963 to 19.3 percent in 1971 of the total U.S. patents issued.[42]
In addition to the import of new technology from abroad, the foreign payments for royalties and management fees provide part of the
financing to fund future research, thus helping the United States to
maintain its leadership role. In 1970, the foreign payments were
equivalent to approximately 11 percent of research and development
spending by all industries in the United States and to nearly 25

percent of the $10.1 billion financed by company rather than federal funds.[43]

While the United States has traded in technology with market economies for many years, it began to trade with the Soviet Union in this area only recently. At the present time, about 49 projects are being developed under the U.S.-USSR Summit Agreement on Cooperation in the Fields of Science and Technology, including four projects involving energy systems.[44] The Soviets lead in some areas of technology because of large research programs that have achieved unique and valuable breakthroughs in metal working, engineering plastics, hydroelectric power, and high-voltage transmission techniques. The United States has already obtained a new low-cost method of extracting magnesium, a new process for cooling blast furnaces, remelting metals, and smelting aluminum, and has purchased Soviet equipment for casting aluminum ingots and for manufacturing thin-walled tubes of hard metal.[45]

The impact of increased technological trade with the Soviet Union is unlikely to be detrimental. Past U.S. experience in transferring automobile technology (delivery of a replica of the Ford Rouge plant in 1934) revealed that the Soviet Union was unable to progress beyond the stage of technology delivered to assume a leadership role. This recorded performance occurred in spite of the fact that the output of trucks and autos of the Soviet Ford factory dominated total Soviet auto and truck production for more than a decade. Gas for the North Star project, on the other hand, would never approach even 5 percent of total Soviet gas production. Assuming Soviet future performance in technology will resemble past performance, it is unlikely that transfer of gas technology to the USSR would threaten U.S. world leadership.[46]

U.S. Considerations in Technology Transfer under North Star

International trade in natural gas is of recent origin. It was made possible by the development of technology involving gas liquefaction and tanker shipments. The British gas industry claims to have played a pioneering role in these developments.[47] Prior to the development of this technology, nations relied on indigenous resources for natural gas. LNG techniques have made it economic to transport gas over very long distances, and supplies can now be drawn from a much wider range of sources than would have been possible with pipelines.

There have also been developments in the technology of wide-diameter pipelines. By 1975, approximately 1,500 million cubic feet

(MCF) of Russian gas per day will be flowing into West Germany, France, and Italy. It will be transported from wide-diameter Russian pipelines through the Austrian network to the transmission systems of Western Europe.[48] By 1980, this flow is scheduled to increase to approximately 2 billion cubic feet per day (BCFD), as indicated in Table 6.6.

In 1973, the Soviet Union transported more than 1 BCFD of gas from the Medvezhe gas deposit, which straddles the Arctic Circle in Western Siberia to the village of Punga, where it entered the Soviet gas distribution system. The pipeline covered a distance of 292 miles and was constructed of 48- and 56-inch pipe.[49]

The first Soviet pipeline in permafrost became operable in 1964 at the Ust-Vilyuy field in the area of the proposed Yakutsk project in Eastern Siberia. It was an experimental pipeline, but has been in operation since that time and continues to supply gas to nearby villages.[50]

The second permafrost pipeline is called the northernmost pipeline in the world and is located on the shores of the Laptev Sea and Arctic Ocean. Gas is transported from the Messoyakha deposit to Norilsk via a 28-inch line for a distance of 150 miles. Norilsk is a major metallurgical center, and the gas is used to fuel the processing furnaces. The pipeline is constructed on supports high enough above the permafrost to keep the heat from melting the frozen ground. It has been said that the surface of the pipe becomes hot enough to fry an egg.

In their extensive operations in permafrost, the Soviets have encountered many problems. In the Messoyakha-Norilsk line, the designers failed to provide for the high winds encountered in the Arctic. Wind friction has caused ruptures in the pipe as it swayed back and forth on supports. A rupture has propagated along the pipe for a distance up to 2 miles. The Soviets are now laying a second pipeline, parallel to the original one.

In drilling operations, the Soviets attempted to use techniques employed successfully in other parts of the country. At the Tazov gas field, precautions were not taken to prevent the permafrost from melting. During drilling, the well collapsed, cratered, and caught fire. So much gas was lost that this field, once ranked among the 10 largest in the Soviet Union, is no longer listed in the Soviet gas reserves.[51]

Soviet managers have described some of the measures being taken to counter the special problems of permafrost operations. When using small-diameter pipelines, the heat generated through compressions is quickly dissipated as the gas returns to ground temperature. In 40-inch lines and larger, it was discovered that the heat continued to rise after compression instead of cooling. To counter this problem, gas-chilling equipment has been installed downstream from the compressors. Research is being conducted involving transmission of gas in a chilled or liquefied state. Russians believe that techniques for pipelining LNG will become practical within the next 10 to 15 years.[52]

The Soviets have begun to install an automatic gas pipeline control system. Gas is introduced into the pipeline, and the temperature, pressure, and flow are monitored and automatically relayed to a control station. Manpower requirements are thereby reduced by 80 percent together with an increase in accuracy and reliability. The system is called "Sever-1" (North-1) and has already been installed on six trunk lines.[53]

There is ample evidence to indicate that world gas technology is well developed, and applications are made throughout the world. There are advantages in exchanging technology through trade because the exchange promotes further development. For example, application was made for U.S. patents involving liquefaction, storage, and barge transportation of natural gas in 1914. A pilot liquefaction plant was built in 1940, and a commercial liquefaction plant was constructed in Cleveland, Ohio, in 1941. The plant operated successfully until 1944, when it was destroyed by fire. The first export of LNG by the United States was accomplished in 1959, when a pilot shipment was made from Louisiana to the United Kingdom. A pilot shipment of LNG from Algeria to Boston occurred in 1968, and the long-term commercial contract for shipment of LNG from Alaska to Japan commenced in 1969.[54]

It was pointed out in Chapter 4 that the first commercial shipments of LNG by a Western government were made from Algeria to Great Britain in 1964, and a total of 40 LNG agreements are in various stages of implementation throughout the world.[55] The history of LNG development shows that the United States probably originated the technology and pioneered in its development, but was overtaken by Western Europe and Japan in commercial application. The U.S. natural gas shortage is forcing the nation back into the forefront of LNG commercial applications.

Although access to American technology would improve permafrost operations and make them more efficient, the Soviets have gained valuable experience from ten years of concentrated operations in this environment. United States access to this information would enhance its operations in the Arctic areas.

As pointed out in Chapter 4, the United States does not have the capability to manufacture wide-diameter pipe and was forced to purchase pipe for the Alyeska pipeline from Japan. If the North Star

project is implemented, assurances have been received that manufacturing capability for wide-diameter pipe would be developed in the
United States.[56] Therefore, transfer of technology to the USSR would
result in some direct and indirect reverse flow of gas technology to
the United States and would contribute to increasing its competitive-
ness in this field.

SUMMARY

Three broad categories of trade authorization factors bear on the
North Star project: (1) the most-favored-nation (MFN) treatment,
(2) the transfer of technology, and (3) the practice of dumping.

Trade authorization factors pose potential obstacles, but they were
found not to be great enough to preclude implementation of North Star:

1. The granting of MFN status to the USSR was found to be of
psychological and political importance and would hinder North Star if
not granted, but would not preclude an agreement by the Soviets. The
economics of the project would not be affected by MFN because the
import of LNG is listed as a free item under U.S. tariff regulations.

2. Transfer of technology is likely to result in administrative delays
since the request for export license is subject to interagency review.
World gas technology is well developed, however, and approval should
be granted because the subject technology is available in Western
Europe and Japan. The implementation of North Star offers a potential
for reverse technology flow in that the United States would develop a
capability to manufacture large-diameter pipe and that the exchange
of information about Soviet operational experience with wide-diameter
pipeline operations in Siberia would be helpful in improving U.S.
Arctic operations.

3. Dumping is not expected to pose problems because the import of
gas under North Star requires prior approval by the FPC; thus, the
problem of dumping is avoided.

NOTES

1. U.S. Tariff Commission, United States East European Trade
Considerations Involved in Granting Most-Favored-Nation Treatment
to the Countries of Eastern Europe, prepared by Anton F. Malish,
Jr., Staff Research Studies No. 4 (Washington, D.C.: U.S. Tariff
Commission, 1972), pp. 1, 5.

2. Commission on International Trade and Investment Policy, United States International Economic Policy in an Interdependent World, 3 vols., commission's report to the president (Washington, D.C.: Government Printing Office, July 1971), vol. 1, pp. 1-10.

3. Council on International Economic Policy, International Economic Report of the President (Washington, D.C.: Government Printing Office, 1974), p. 3.

4. Commission on International Trade and Investment Policy, op. cit., vol. 1, p. 1.

5. Franklyn D. Holzman, "East-West Trade and Investment Policy Issues," in International Economic Policy, commission's report, op. cit., vol. 2, pp. 117-18.

6. Mikhail V. Condoide, Russian American Trade (Columbus: Ohio State University, 1946), pp. 117-18.

7. Edward T. Wilson, et al., "U.S.-Soviet Commercial Relations," in U.S. Congress, Joint Economic Committee, Soviet Economic Prospects for the Seventies (1973), p. 639.

8. Max M. Laserson, The American Impact on Russia—Diplomatic and Ideological—1784-1917 (New York: Macmillan Company, 1950), p. 46.

9. Condoide, op. cit., pp. 118-19.

10. Ibid., pp. 120-28.

11. Ibid., pp. 121-22

12. Laserson, op. cit., pp. 353-58.

13. Condoide, op. cit., p. 91.

14. Ibid., p. 129.

15. Ibid., pp. 135-45.

16. U.S. Tariff Commission, East European Trade, op. cit., p. 30.

17. Ibid., pp. 5-7.

18. U.S. Department of Commerce, U.S.-Soviet Commercial Agreements, 1972: Texts, Summaries and Supporting Papers (Washington, D.C.: Government Printing Office, 1973), pp. 76, 88-91.

19. U.S. Tariff Commission, Impact of Granting Most-Favored-Nation Treatment to the Countries of Eastern Europe and the People's Republic of China, prepared by John E. Jelacic, Staff Research Studies No. 6 (Washington, D.C.: U.S. Tariff Commission, 1974), pp. 20, 55.

20. Ibid., pp. 23-25.

21. Ibid., pp. 11-13.

22. Ibid., pp. 19-20.

23. George S. Schukin (chairman, Kama River Purchasing Commission), "The Soviet Position on Trade with the United States," Columbia Journal of World Business 8, no. 4 (winter 1973): 48-50.

24. Stephen Isaacs, "End Seen to Trade Impasse," Washington Post, September 7, 1974; and press release from Senator Henry M. Jackson, January 26, 1975.

25. U.S. Tariff Commission, Most Favored Nation Treatment, op. cit., p. 4.

26. U.S. Tariff Commission, Tariff Schedules of the United States Annotated (1972), T.C. Pubn. 452 (Washington, D.C.: Government Printing Office, 1972), p. 276.

27. Holzman, op. cit., pp. 368-69.

28. Ibid., p. 369.

29. U.S. Tariff Commission, East European Trade, op. cit., p. 44; and U.S. Tariff Commission, Pig Iron from East Germany, Czechoslovakia, Romania, and the USSR, TC Pubn. 265 (Washington, D.C.: U.S. Tariff Commission, September 1968).

30. U.S. Tariff Commission, Titanium Sponge from the USSR, TC Pubn. 255 (Washington, D.C.: U.S. Tariff Commission, July 1968).

31. Council on International Economic Policy, op. cit., pp. 3-6.

32. U.S. Department of Commerce, op. cit., p. 76.

33. U.S. Statutes at Large 83, 841-42 (1969), as amended; U.S. Code, Supp. 2, Title 50, App. 2402 (1970); and P.L. 92-412, Title 1, Sec. 103, August 29, 1972; Statutes at Large 86, 644 (1972).

34. P.L. 93-618, 93rd Cong., H.R. 10710, Trade Act of 1974, January 3, 1975.

35. U.S. Statutes at Large, loc. cit., U.S. Code, loc. cit.; and P.L. 92-412, loc. cit.

36. P.L. 93-618, loc. cit., pp. 88-89.

37. William J. Casey, testimony, House Subcommittee Hearings on International Economic Policy, 1974, p. 660.

38. U.S. Department of Commerce, East-West Trade, op. cit., p. 59.

39. Dr. Roger E. Shields, deputy assistant secretary of defense for international economic affairs, "National Security Impact of U.S. Capital Investment and High Technology Transfers to the USSR," testimony at hearings before the Subcommittee on Multinational Corporations, Senate Committee on Foreign Relations, July 18, 1974.

40. Ibid.

41. Council on International Economic Policy, op. cit., p. 70.

42. Ibid., p. 72.

43. J. Fred Bucy, testimony, in U.S. Congress, House Committee on Science and Astronautics, The Technology Balance, 93rd Cong., 1st and 2nd sess., December 1973, p. 63.

44. Steven Lazarus, Testimony in U.S. Congress, Joint Economic Committee, Soviet Economic Outlook, Hearings, 93rd Cong., 1st sess., July 19, 1973, p. 14.

45. John K. Tabor, undersecretary, Department of Commerce, "Achieving a Favorable Balance," International Journal of Research Management (July 1974): 9-10.

46. Frederick B. Dent, secretary of commerce, testimony given before the Subcommittee on Multinational Corporations, Senate Committee on Foreign Relations, July 17, 1974.

47. J. V. Licence, ''Siberia in the Context of World Natural Gas Supplies,'' paper presented for NATO Round Table Meeting, Brussels, January 30, 1974, pp. 1-2.

48. Ibid.

49. Robert E. Ebel, ''Gas in the Soviet Union,'' paper presented at the 48th Annual Fall Meeting of the Society of Petroleum Engineers at Las Vegas, Nevada, October 1-3, 1973, pp. 2-3.

50. Robert E. Ebel, ''Russians 'Think Big' in Future Pipe Line Plans,'' Pipe Line Industry, November 1972, p. 30.

51. Ebel, ''Gas in the Soviet Union,'' op. cit., p. 2.

52. A. M. Zinevitch, Avenosov, ''Soviet Construction Methods for Large Diameter Lines,'' Pipe Line Industry, February 1973, pp. 42, 44.

53. O. Buzuluk, ''Sever-1'' [North-1], Krasnaya Zvezda [Red Star], July 15, 1973, p. 4.

54. U.S. Federal Power Commission, Natural Gas Supply and Demand, 1971-90, Staff Report No. 2 (Washington, D.C.: Bureau of Natural Gas, February 1972).

55. Dean Hale, ''International LNG Movements Continue to Expand,'' Pipeline and Gas Journal (December 1973): 61-63.

56. Jack H. Ray, interviews and correspondence, September 1973 to July 1974.

10

TRADE FINANCING
FACTORS

Almost all large-scale world trade is conducted on a credit basis. The very large quantities of money needed to support large-scale commerce are not readily available. Big business simply is not done "on a cash basis."[1]

The North Star project is large-scale commerce and requires financing arrangements to cover some $3.4 billion in exports of goods and services from the United States to the USSR. American planners desire to raise the capital required in the U.S. domestic capital markets, hopefully through arrangements that would not place undue pressure on the U.S. balance-of-payments structure.[2]

The United States has been in a chronic balance-of-payments deficit since 1958, and the energy crisis has aggravated the imbalance since 1973. More was paid for petroleum imports in July 1974 ($2.3 billion) than in any other prior month on record. The $13.4 billion paid for petroleum in the first seven months of 1974 were almost twice the bill for the entire year of 1973. The trade deficit for July amounted to $728.4 million, and Commerce Secretary Frederick B. Dent identified the import of petroleum as the major culprit in the trade deficit.[3]

The North Star project was originally planned to be entirely financed in the U.S. domestic capital market because the U.S. balance-of-payments deficit would be aggravated if foreign capital markets were used. Foreign government financing ties loan arrangements to the purchase of foreign equipment. Upon delivery of Soviet gas, the United States would be required to pay in cash for that portion of gas imports produced by the foreign-financed equipment.[4]

The attempt to finance North Star through the American domestic capital markets encounters three general categories of obstacles: (1) restrictions imposed by Export-Import Bank financing, (2) lending limitations associated with commercial bank financing arrangements, and (3) legal limitations related to debt default. Each of these

categories is analyzed separately as it applies to North Star financing
arrangements.

EXPORT-IMPORT BANK FINANCING

In fiscal year 1973, U.S. exports totaled $57.9 billion; the Export-
Import Bank supported about $10.5 billion of this trade through loan
or loan guarantees.[5] Approximately 18 percent of all exports received
some type of financing assistance from the bank. In fiscal year 1974,
financing assistance provided by the bank has increased to approximately $13 billion.[6]

The Export-Import Bank operates as an independent agency of the
U.S. government. Legislation removed it from operation of the U.S.
budget on August 17, 1971. It does not receive any appropriated funds
or other tax revenues. The bank funds its operations by borrowing as
needed from the U.S. Treasury on a daily basis, much as a commercial
bank does. It buys or sells money in the federal funds market and then
repays these Treasury borrowings from the receipts of its operations
and the proceeds of private market borrowings. The net draw on the
U.S. Treasury for the fiscal year as a whole is virtually zero.[7]

In support of the policy to expand trade, Congress has directed the
Export-Import Bank to provide loans and loan guarantees to American
exporters that would be competitive with those provided by foreign
governments to competitors.

Until February 1974, the bank carried out this mandate by providing
loans at 6 percent interest to cover 45 percent of the transaction cost.
The remaining funds were provided by a 10 percent cash payment and
45 percent commercial bank financing at market rates of interest.[8]

The rising interest rates in the U.S. domestic capital markets
forced the Export-Import Bank to raise its interest rate to 7 percent
in February 1974. Beginning in April, the bank decreased its participation rate from 45 percent to as low as 25 percent on some loans when
loan demands continued to increase and the bank was approaching its
authorized limit in loanable funds. The bank continued to conduct
profitable operations, however, because the bank's weighted cost of
borrowing was below 6 percent until late in 1973 and did not reach 6.8
percent until April 1974. The weighted cost of borrowing exceeded 7
percent for the first time in May 1974. The bank announced that it
was abandoning the fixed interest rate on July 9, 1974. In the future,
an interest rate would be set between 7 and 8.5 percent, appropriate
to each transaction. The interest rate would depend on a combination
of factors: size of loan, maturity, competitive situation, and participation rate. The reasons given for the changes were the increased

cost of borrowing and inadequate resources to finance all the loans requested.[9]

The bank earned a profit of $107 million in fiscal year 1974 and paid a $50-million dividend to the Treasury. It was the twenty-fourth consecutive dividend for a total contribution of $855 million over the period.[10]

Nevertheless, the bank found itself unable to meet the heavy loan demand of American business. At the end of fiscal year 1974, the bank had about $700 million in loans that were analyzed and approved but had to be carried over into 1975. Under the annual ceiling established by Congress, there were insufficient funds to cover the loans in the current year. Large projects, which are important for their contribution to the U.S. balance-of-payments posture, such as nuclear reactors, airplanes, and LNG ships, are delayed by the sheer unavailability of capital.[11]

During fiscal year 1973, transportation items accounted for 14 percent and electric and nuclear power plants 9 percent of all loans made by the bank.[12] There are estimated to be 29 nuclear power plants in various stages of development around the world, with many more to come. These will require from $3 to $5 billion in financing between 1974 and 1978. In order to support the markets established abroad by American businessmen, an increase in the authorized ceiling, from $20 to $30 billion, has been requested from Congress.[13] Congress approved an increase in authorized ceiling to $25 billion, and it was enacted into law on January 4, 1975.

In order to support U.S.-Soviet trade, a Guarantee Agreement was signed between the Export-Import Bank of the United States and the Ministry of Foreign Trade of the USSR on March 21, 1973.[14] Under this agreement, the government of the USSR agreed to guarantee repayment to the Export-Import Bank of any credits extended by the bank to the Bank of Foreign Trade of the USSR.

The first credits extended under the agreement were approved in the amount of $225 million with the signing of the agreement.[15] The terms of the loans were similar to those provided for American exports elsewhere in the world. The Soviets were to pay 10 percent cash, would receive an Export-Import Bank loan to cover 45 percent of the transaction, and had to arrange for a 45 percent commercial bank loan, without an Export-Import Bank guarantee of repayment.

Fifteen transactions with the Soviet Union were approved through May 21, 1974. A list of the individual projects and loan totals is found in Table 10.1. Total credits in the amount of $470 million have been approved to support U.S. exports in the amount of $1,044 billion to the Soviet Union. In each of these transactions, repayment of the Export-Import Bank credits is guaranteed by the Soviet government.

A temporary halt in granting Export-Import Bank loans to the USSR
and nations of Eastern Europe occurred in March 1974. The loans
were suspended on March 11, 1974, on the basis of an opinion of the
comptroller general of the United States stating that each individual
transaction involving these countries was subject to a finding by the
president that such a transaction was in the national interest.[16] In
response, the general counsel of the Export-Import Bank prepared an
opinion justifying continued loans to these countries which was forwarded to the president.[17] The attorney general of the United States
likewise submitted a concurring opinion to the president on March 21,
1974, which affirmed prior bank practice. The issuing of a blanket
presidential determination for each country was found to be consistent
with existing law. On the basis of this opinion, the Export-Import Bank
resumed financing transactions with the USSR and countries of Eastern
Europe on March 22, 1974.[18]

Certain members of Congress strongly opposed the resumption of
credit to Communist nations. Senator Richard B. Schweiker, of
Pennsylvania, introduced legislation to prohibit U.S.-government-
supported investments in Soviet energy projects. He proposed an
amendment to the second supplemental appropriation bill to prohibit
the Export-Import Bank from expending any funds until the bank com-
plies with the comptroller general's opinion.[19]

The restrictive provisions, which eventually were enacted into law
by the Export-Import Bank Amendments of 1974, included (1) a $300-
million total limit including a $40-million sublimit for energy projects
on loans to the Soviet Union, (2) a requirement for a separate determination of national interest for each transaction in which the Export-
Import Bank would loan $50 million or more, and (3) a 25-day advance
notification to Congress for any loan that exceeds $60 million and for
any loan covering an energy project in the Soviet Union that exceeds
$25 million. The above limits may be increased if the president
determines that a higher limit is in the national interest and if the
Congress approves his determination by adopting a concurrent resolution.[20] As of June 30, 1974, the financial resources of the bank
were being stretched to their limits. The charge against the bank's
authorized $20-billion commitment authority was $17.6 billion, and a
$5.6-billion loan authorization was requested for fiscal year 1975.
Planned loan operations would carry the bank $800 million over the
$20-billion authorization during 1975.[21]

The Export-Import Bank has not received a loan application for
North Star, and the necessary approval for the project has not been
received from the Federal Power Commission (FPC). If application
for a loan were received, the project could not be financed under the
current loan authorization of the bank for 1975. The reasons include

TABLE 10.1

Approved Export-Import Bank Credits to Promote Soviet Trade
(in U.S. dollars, as of July 1, 1974)

Items Exported	Total Cost of Transaction	Cash Payment	Export-Import Bank Loan	Commercial Bank Unguaranteed Loan	Participating Commercial Bank	Terms in Years
Tableware and dishware plant	6,893,138	689,314	3,101,912	3,101,912	Wells Fargo (San Francisco)	10
500 Submersible electrical pump units	25,937,000	2,593,700	11,671,650	11,671,650	French Amer. Bk Corp. et al.	7
Kama River truck plant	192,211,000	19,211,000	86,450,000	86,450,000	Chase Manhattan et al.	12
250 circular knitting machines	5,620,000	562,000	2,529,000	2,529,000	Bankers Trust Co.	7
Tableware plant	21,833,000	2,183,300	9,824,850	9,824,850	Bankers Trust Co.	10
Piston mfg. machinery	14,358,118	1,435,812	6,461,153	6,461,153	NA	8
Iron ore pellet plant	36,000,000	3,600,000	16,200,000	16,200,000	NA.	8

38 gas reinjection compressors	26,251,565	2,625,157	11,813,204	11,813,204	Wells Fargo (San Francisco)	7
Piston mfg. transfer line	15,722,042	1,572,204	7,074,919	7,074,919	French Amer. Bk Corp.	8
Equipment to manufacture drums (tractor)	6,000,000	600,000	2,700,000	2,700,000	Franklin Natl Bk	8
Kama River truck plant	149,900,111	15,000,111	67,500,000	67,500,000	Chase Manhattan et al.	12
Flywheel transfer lines	7,458,100	745,810	3,356,145	3,356,145	NA	7
Acetic acid plant	44,515,000	4,451,500	20,031,750	20,031,750	NA	10
Canal lining material	6,600,000	660,000	2,970,000	2,970,000	NA	5
Ammonia urea plants	400,000,000	40,000,000	180,000,000	180,000,000	Bk of America et al.	12
Total	1,043,999,074	104,399,908	469,799,583	469,799,583		

Sources: Export-Import Bank of the United States, press releases of March 21, September 27, and December 21, 1973, and January 18, February 22, March 22, and May 21, 1974.

inadequate funds, the bank's responsibility to support markets already established, and the fact that investing so heavily in one country would place the bank's investment portfolio out of balance.[22] The following statement was made in recent testimony:

> . . . If application were made on these projects, we would not be able to handle them, or any major part of them, under the proposed loan limitation for 1975 without impairing our ability to finance exports in established markets all around the world to a degree which we would not be willing to do. . . . We must continue to support exporters who work on those markets on a worldwide plan.
>
> We could not afford to tilt that heavily in the direction of a single country. So, if it were desirable to finance the projects of this type, Congress would have to increase our loan limit and Eximbank's directors would have to satisfy themselves that there would be no adverse consequences to our domestic economy.[23]

The financing of North Star through Export-Import Bank loans would accrue increased costs because of the new flexible interest arrangements instituted in July 1974. The interest rate has been increased from 7 percent to a variable rate between 7 and 8.5 percent. Also, under recent changes made in loan procedures, the participation rate by the Export-Import Bank may drop below 45 percent of total cost. The net effect of these changes is to increase the average effective interest rate for borrowing under the project. Under the 7 percent interest rate and 45 percent participation, the average effective rate of financing was approximately 8 percent. This rate was higher than that offered by competitors, except West Germany. For example, Japan granted loans as low as 5.5 percent, and France 6.35 percent, and foreign government participation was as high as 90 percent; the average effective rate of interest was therefore below that offered by the Export-Import Bank.[24]

Financing of North Star through the Export-Import Bank is possible but unlikely because of considerable political opposition to granting loans to the Soviet Union. Widespread support for the loan restrictions was generated because of a misunderstanding concerning the apparently concessionary 7 percent interest rate on Soviet loans compared to the rate charged in the U.S. domestic capital market. International financing arrangements are complex, and the average layman does not understand the two-tier interest rate market which exists in international finance. Open-market interest rates on business loans run as high as 14 percent, but the government-subsidized

market grants loans designed at a lower rate to encourage exports.
These loans are tied to export purchases and can be obtained at a
rate below 7 percent.[25] Any businessman can obtain the lower rate
provided that specified products are purchased from foreign producers. The interest subsidy is designed to promote efforts, thereby
creating jobs for the economy. Hopefully, the change to flexible
interest rates by the Export-Import Bank will tend to equalize interest
rates, thereby reducing some of the opposition to the granting of
credits to the Soviet Union. In financing for North Star, however,
specific congressional authorization will be needed to provide the
amount of funds required to finance the project, because the approved
authorization limit for loans by the Export-Import Bank is inadequate
to support a project as large as North Star.

COMMERCIAL BANK FINANCING

Legal limitations and judgmental decisions concerning prudent
risk form the obstacles for financing the North Star project by
commercial banks. United States law limits the amount any bank
can lend to a single person to 10 percent of the bank's net worth.
The legal limits are set forth in Title 12, United States Code, Section
84:

> The total obligations to any national banking association of
> any person, copartnership, association, or corporation shall
> at no time exceed 10 per centum of the amount of the capital
> stock of such association actually paid in and unimpaired
> and 10 per centum of its unimpaired surplus fund.[26]

Since the Soviet government owns all corporations and banking
institutions, does U.S. law limit the total of all loans to 10 percent of
a lending bank's net worth? The legal lending limits have not yet
been reached by the Soviet Union; so a positive answer in terms of a
court decision has not been rendered. The legal answer will depend
on specific details in each case.

Difficulties arise when state-owned enterprises attempt to borrow
money in U.S. financial markets. In the application of U.S. law, the
concept of one person (the borrower subject to the 10 percent limit
factor) has been broadened. If an institutional borrower can demonstrate a capability of generating its own funds sufficient to repay the
loan, it can be considered an independent person, even though it is
owned by a government parent. Export-Import Bank loan approval is

granted on a case-by-case basis, after analysis of a borrower's pay-back capability. For example, a loan was approved for the government-owned PeMex Corporation of Mexico even though the loan exceeded the 10 percent limit on all obligations outstanding by the lending bank to the government of Mexico.[27]

Under this interpretation, North Star appears to qualify as "one person" since the debt would be self-liquidating. But, theoretically, North Star could be broken down into an "independent" pipeline division and a liquefaction division, each with its own capability to generate funds, thereby doubling its borrowing capacity. In this case, however, adequate financial data would have to be submitted by the Soviets to verify a finding of self-generating funds capability sufficient to repay the loans. In the past, the Soviet Union has been reluctant to provide such data.

The question of how much financing can be provided for North Star by commercial banks arises. Not all U.S. banks are anxious or even willing to loan funds to the Soviets, but the larger banks are willing to accept this risk. If one assumes that the 50 largest U.S. banks would participate, the legal lending limit to one person would be $1.9 billion. The combined net worth of the 50 largest U.S. commercial banks is $19,159,959,000.[28] Data on the individual net worth of each of the banks are provided in Table 10.2. The total Export-Import Bank participation loans outstanding to the Soviet Union by U.S. commercial banks is $469,799,583, as shown in Table 10.1. The above data reveal that U.S. banks would have to tie up a considerable portion of their lending capacity if they were to provide financing for North Star.

The judgmental decision on whether to finance North Star confronts the bankers with the problem of prudent risk. Discussions with bankers in the New York and Washington, D.C., areas reveal a reluctance to commit large loans to a single purchaser for a long period of time, regardless of how good a reputation the borrower demonstrates. Loans beyond 10 percent of a bank's net worth are considered to stretch the policy of prudent risk. A New York banker stated that a way would be found to finance a project, however, if analysis showed it to be economically desirable.[29] Regular and unannounced audits are conducted of a bank's accounts including its loan portfolio, and most bankers prefer not to receive comments about the "concentration of credits" in their postaudit reports.[30]

The North Star project currently requires financing in the amount of approximately $3.5 billion. If the Export-Import Bank loans and guarantees are granted to cover $2 billion, the remaining funds could be generated through commercial bank financing. Failure of the Export-Import Bank to participate in North Star would require banks to exceed current practices concerning prudent risk. It is possible for U.S. banks to make loans through wholly owned foreign subsidiaries because foreign subsidiaries are not subject to U.S. legal limitations on lending.[31] However, it is believed that banks would consider the risk to be too great to justify the undertaking. A more logical course of action would be to form a financing consortium in which foreign banks would share the financial risk. Under this arrangement, foreign banks could replace the Export-Import Bank and share in the financial risk and rewards of North Star.

A precedent for North Star exists in the banking consortium which was formed to finance Algerian gas exports, and the Japanese have proposed a mixed international financing arrangement for the Yakutsk project in Eastern Siberia. Initially, the Japanese proposed to finance the entire project with Japanese capital, but the attitude of Japanese businessmen changed markedly in 1972. The protocol signed by Soviet and Japanese officials on April 22, 1974, indicating preliminary agreement on three major Siberian developments, is conditioned on U.S. participation for final approval.[32] A wish to reduce political risk and diversify economic risk was the primary reason for the Japanese change in attitude.

The financing of North Star through a consortium of international banks has been investigated by American negotiators and found to be feasible. Foreign participation would require that part of the equipment be purchased from foreign sources.[33] There would be disadvantages in the sense that the U.S. balance of payments would be affected negatively, and the project would be delayed by approximately 18 months.[34] An advantage of consortium financing would be the diversification of economic risk. Lower U.S. financing would be required, thus reducing U.S. exposure. Also, political risk would be reduced in that the Soviet Union would be reluctant to renege on gas deliveries because such action would result in default on loans from all participating nations.

FINANCING LIMITATIONS RELATED TO DEBT DEFAULT

Financing limitations for the North Star project are posed by Soviet debts to the United States, which are considered to be in default from World War I and World War II.

The Johnson Debt Default Act of 1934 makes it a criminal offense for any individual in the United States to sell or purchase bonds, securities, or other financial obligation of any foreign government that is in default on obligations to the United States. Also, the act prohibits the making of any loan on behalf of such government.[35]

TABLE 10.2

Net Worth of the Fifty Largest U.S. Commercial Banking Companies
(in thousands of U.S. dollars)

Company	Net Worth	Assets
1. First National City Corp.	1,816,641	44,019,218
2. Bank America Corp.	1,550,648	49,404,764
3. Chase Manhattan Corp.	1,348,238	36,790,909
4. Morgan & Co.	956,841	20,374,529
5. Manufacturer's Hanover Corp.	794,606	19,850,398
6. First Chicago Corp.	698,914	15,558,497
7. Western Bancorp.	688,956	17,902,598
8. Chemical New York Corp.	687,726	18,592,219
9. Continental Ill. Corp.	676,725	16,870,180
10. Mellon Natl. Corp.	555,722	9,600,499
11. Security Pacific Corp.	550,911	13,478,783
12. Bankers Trust N.Y. Corp.	540,755	18,514,550
13. First Natl. Boston Corp.	432,953	8,004,705
14. Marine Midland Bks. Inc.	407,068	13,044,315
15. Wells Fargo & Co.	392,774	11,767,725
16. First Bk. System Inc.	386,553	6,514,059
17. Northwest Bancorp.	361,582	6,517,421
18. National Detroit Corp.	359,824	6,711,761
19. Cleveland Trust Co.	319,023	3,464,902
20. Charter New York Corp.	319,023	9,738,989
21. Croker National Corp.	314,369	9,767,969
22. Wachovia Corp.	266,522	3,915,256
23. First Pennsylvania Corp.	263,221	6,192,557
24. First Intl. Bancshares Inc.	249,432	5,052,272
25. Bank of New York Co.	218,163	3,992,105

Company	Net Worth	Assets
26. Philadelphia National Corp.	201,855	4,029,655
27. Citizens & So. Natl. Bank	197,855	3,089,234
28. Seattle First Natl. Bank	195,486	3,969,238
29. Natl. Bank of No. America	192,680	2,887,146
30. Pittsburg Natl. Corp.	189,926	2,744,321
31. Harris Bankcorp.	188,294	3,782,388
32. BancOhio Corp.	186,435	3,080,657
33. Detroit Bank Corp.	181,271	2,908,676
34. NCNB Corp.	178,062	4,112,136
35. Notrust Corp.	173,334	3,462,842
36. Union Bancorp.	168,002	4,849,625
37. Tex Commerce Bkshares	164,074	2,828,639
38. Republic Natl. Bk. of Dallas	157,746	4,214,793
39. First Wisc. Bkshares Corp.	155,948	3,431,133
40. Franklin N.Y. Corp.	152,013	5,006,616
41. First City Bancorp. of Tx.	150,902	3,580,643
42. Southeast Banking Corp.	144,598	2,497,163
43. Girard Co.	144,368	3,353,191
44. U.S. Bancorp.	140,143	2,725,656
45. Michigan National Corp.	135,764	2,892,143
46. Lincoln First Banks Inc.	132,514	2,555,678
47. Fidelity Corp. of Penna.	132,263	2,628,935
48. Valley Natl. Bk. of Arizona	130,286	2,959,354
49. Manufacturers Natl. Corp.	128,087	2,740,803
50. BanCal Tri-State Corp.	91,803	3,055,522
Total	19,159,959	459,027,367

Source: Compiled from Eleanor Johnson Tracy, ''The Fortune Directory,'' Fortune (July 1974): 112-15.

The text of the criminal code, as related to the Johnson Debt Default Act of 1934, states:

> Whoever, within the United States, purchases or sells the bonds, securities, or other obligations of any foreign government or political subdivision thereof or any organization or association acting for or on behalf of a foreign government or political subdivision thereof, issued after April 13, 1934, or makes any loan to such foreign government, political subdivision, organization or association, except a renewal or adjustment of existing indebtedness, while such government, political subdivision, organization or association, is in default in the payment of its obligations, or any part thereof, to the United States, shall be fined not more than $10,000 or imprisoned for not more than five years, or both.
>   This section is applicable to individuals, partnerships, corporations, or associations other than public corporations created by or pursuant to special authorizations of Congress, or corporations in which the United States has or exercises a controlling interest through stock ownership or otherwise. While any foreign government is a member both of the International Monetary Fund and of the International Bank for Reconstruction and Development, this section shall not apply to the sale or purchase of bonds, securities, or other obligations of such government or any political subdivision thereof or of any organization or association acting for or on behalf of such government or political subdivision, or to making of any loan to such government, political subdivision, organization, or association.[36]

The act was originally intended to protect American investors against the danger of loss through the purchase of bonds and other obligations of countries that were likely to default on debts. In support of this interpretation, Attorney General Homer Cummings issued a formal ruling shortly after the legislation was enacted in 1934 stating that the legal restrictions of the act were intended to include debt instruments and not financial instruments associated with trade or other commercial transactions. The act prohibited the sale of instruments

> . . . which had been sold to the American public to raise money for the use of foreign governments issuing them. . . .
>   . . . It was obviously not the purpose of Congress to discontinue all commercial relations with defaulting countries.[37]

The ruling went on to specify that trade financing transactions such as the acceptance of time drafts were not prohibited. Trade financing transactions were legal, provided they were "conducted in good faith and not as mere subterfuges to circumvent" the purpose of the act.

The question of denial of credit loans to the Soviet Union because of debt default arose in 1963. The USSR had requested credits to purchase U.S. grains due to a Soviet crop failure. Attorney General Robert F. Kennedy issued a formal opinion clearly distinguishing between public debt securities, which are prohibited, and trade financing transactions, which are not prohibited. Financial obligations issued in the normal course of trade would not violate the act. These were identified as:

> . . . credit transactions in which private exporters commonly engage in connection with export sales on credit, involving the assignment of negotiation of contract rights or commercial paper.[38]

An opinion by Attorney General Kennedy stated:

> Direct recourse to the legislative history of the act confirms that both distinctions here made—that between loans and commercial credit, and between securities and commercial paper— reflect accurately the intention of Congress and the policy it sought to implement. As noted by Attorney General Cummings, it was obviously not the purpose of Congress to interfere with the ordinary incidents of trade relations with the defaulting nations as distinguished from participation by them in the capital markets of the United States.[39]

The act and interpretation of the act through formal opinion of the attorney general clearly distinguish between trade financing, which is legal, and investment financing, which is prohibited. In the North Star project, the technical, legal definition of investment would categorize the project as barter trade rather than investment. Equity is not passed from the USSR to the United States. The large outlay of funds and the long-term repayment in gas does, however, convey an investment impression.

In 1967, Acting Secretary of State Nicholas Katzenbach requested an opinion from the attorney general on whether the Johnson Act was applicable to certain financing transactions on projects in the USSR and Eastern Europe. Acting Secretary Katzenbach specifically mentioned that "export sales to some of the countries in question may have to be financed through receipt of payment from foreign buyers in kind rather than currency."[40]

In a formal opinion, Attorney General Ramsey Clark cited earlier opinions by Cummings and Kennedy, reviewed committee debates in both houses of Congress, and then ruled that, in his opinion, the proposed financing transactions would be legal under the act.[41]

The reasoning behind this opinion by Attorney General Clark was as follows:

> . . . Although the cited opinion did not deal with the three forms of financing about which Acting Secretary Katzenbach explicitly inquires—lines of bank credit, barter arrangements, and deferrals of payment pending development of earnings—I can discern no valid ground of distinction, from the standpoint of applicability of the Johnson Act, between these forms of financing and the ones which in that opinion were found to be generally permissible under the act. The reasoning of the opinion supports the general conclusion that financing arrangements lie beyond the scope of the act if they are directly tied to specific export transactions, if their terms are based upon bona fide business considerations, and if the obligations to which they give rise move exclusively within the relatively restricted channels of banking and commercial credit.[42]

Formal precedent concerning the applicability of the Johnson Debt Default Act is established by commercial practice dating back to the 1930s and by formal opinions of three different individuals who have held the office of attorney general. It appears clear that the act applies to the floating of public debt securities by foreign governments in the U.S. domestic capital market, which type of financing is illegal under the act. Long-term financing of American exports, however, has been conducted for many years and appears legal under the act. At least, this is true by implication since no criminal indictment has ever been brought under the statute. Nevertheless, the Johnson Act applies negative pressure on proposals from nations in default status. Investors have either requested clarification from the attorney general or avoided the problem by financing projects through third countries.[43] This practice is beyond the reach of the law because it applies "within the United States."[44]

As the North Star project proposes to finance the export of U.S. equipment and services through a barter exchange for Soviet natural gas, it appears to be legal under the Johnson Debt Default Act of 1934 on the basis of precedent.

A second obstacle posing financing limitations for North Star concerns Soviet World War II debts. Soviet obligations to the United States

under the Lend-Lease agreement of World War II were a deterrent to
Soviet-American commercial relations since the end of the war.
Repeated attempts to resolve the issue ended in failure. However,
the Lend-Lease Settlement Agreement was finally signed, on October
18, 1972.[45]

Under the terms of the agreement, the Soviets agreed to pay an
amount of at least $722 million over the period ending July 1, 2001.
Three payments would be made prior to July 1, 1975. Repayment of
the remainder of the debt would commence after the United States had
granted most-favored-nation (MFN) status to the Soviet Union.[46]

Should the United States fail to grant MFN status to the Soviet
Union, the agreement permits the Soviets to suspend debt repayment
until such time as MFN is granted without falling into default on the
debt. The restrictions that Congress has included in the Trade Act
of 1974 and in the Export-Import Bank Amendments of 1974 appear
to withhold MFN status intentionally from the Soviet Union. Section
404 of the Trade Act explicitly denies the granting of MFN treatment
to a country which is in arrears on its obligations under a Lend-Lease
Settlement Agreement. Yet, if the Soviet Union makes its third payment by July 1, 1975, it could withhold further payment and would not
be in technical default until the United States granted MFN status. In
this case, the executive and legislative branches appear to be working
against each other.

SUMMARY

Financing of North Star through the American domestic capital
market encounters three general categories of obstacles: (1) Export-
Import Bank financing, (2) loan limitations on commercial bank
financing, and (3) debt default limitations.

1. Export-Import Bank financing is legally possible. The president
would have to make a determination that the North Star project was
in the national interest and that determination approved by a concurrent resolution of the Congress. Then specific authorizing legislation
is required to make the needed funds available. Strong political opposition to this action exists on the part of certain legislators.

2. If an Export-Import Bank loan or loan guarantees were provided
to cover approximately one-half of the project cost, the remaining
funds could be generated through commercial bank financing arrangements. Although complete financing through commercial bank financing
is possible, such action is not likely to occur. Failure of the Export-
Import Bank to participate in the project would require commercial

banks to exceed current practices concerning prudent risk. An alternative course of action would be to form a consortium in which foreign banks would share financial risks of the project.

3. Legal limitations of the Johnson Debt Default Act do not appear to apply to North Star because the project involves financing of U.S. exports rather than the selling of securities in the American domestic capital market. The North Star proposal appears to be legal under the act on the basis of precedent and three formal opinions by the attorney general of the United States. The problem of World War II debt default was resolved upon signing of the Lend-Lease Settlement Agreement on October 18, 1972, and the Soviet Union will not be in default provided the third debt payment is made on July 1, 1975. Future debt payments by the USSR are dependent upon the granting of MFN status by the United States.

NOTES

1. William J. Casey, Export-Import Bank press release, July 9, 1974, p. 25.

2. Jack H. Ray, "The North Star Soviet Liquefied Natural Gas (LNG) Import Project," statement in U.S. Congress, House, Committee on Banking and Currency, p. 216.

3. "Trade Deficit Deepens," Plain Dealer, August 27, 1974, p. 5c.

4. Ray, op. cit.

5. Export-Import Bank of the United States, Eximbank and the World of Exports, statement of condition, fiscal year 1973 (Washington, D.C.: Government Printing Office, 1974), pp. 19, 28-29.

6. Casey, Export-Import Bank press release, July 9, 1974, p. 2.

7. Export-Import Bank of the United States, op. cit., p. 24.

8. William J. Casey, testimony, House Subcommittee Hearings on International Economic Policy, 1974, pp. 655, 656.

9. Casey, Export-Import Bank press release, July 9, 1974, pp. 2-9.

10. Ibid., pp. 3-4.

11. Ibid., pp. 4, 14.

12. Export-Import Bank of the United States, op. cit., pp. 28-30.

13. Henry Kearns, testimony in U.S. Congress, Senate, Committee on Banking, Housing, and Urban Affairs, Export-Import Bank of the United States. Hearings before the Subcommittee on International Finance on S. 1890, 93rd Cong., 1st sess., 1973, p. 18.

14. Export-Import Bank of the United States, press release, March 21, 1973.

15. Ibid.

16. U.S. Senate, Committee on Banking, Housing and Urban Affairs, The Role of the Export-Import Bank and Export Controls in U.S. International Policy. Hearings before the Subcommittee on International Finance on S. 1890 and S. 3282, 93rd Cong., 2nd sess., 1974, pp. 119-22. (Cited hereafter as Senate Subcommittee Hearings on the Role of the Export-Import Bank, 1974.)

17. Ibid., p. 114; and Export-Import Bank of the United States, press release, March 22, 1974.

18. Senate Subcommittee Hearings on the Role of the Export-Import Bank, 1974, pp. 114, 119-22.

19. Senate Subcommittee Hearings on the Role of the Export-Import Bank, 1974, p. 118.

20. P.L. 96-646, Export-Import Bank Amendments of 1974, pp. 7380-81.

21. Senate Subcommittee Hearings on the Role of the Export-Import Bank, 1974, p. 39.

22. Casey, testimony, in House Subcommittee Hearings on International Economic Policy, 1974, p. 662.

23. Ibid.

24. Casey, testimony, in Senate Subcommittee Hearings on the Role of the Export-Import Bank, 1974, pp. 42-44.

25. Casey, Export-Import Bank press release, July 9, 1974, p. 14.

26. United States Code Annotated, Title 12, Banks and Banking, Secs. 1-530 (St. Paul, Minn.: West Publishing Co., 1972), p. 305.

27. Interview with Robert A. Mullin, Department of the Treasury, Office of the Comptroller of the Currency, Washington, D.C., during August 1974.

28. Eleanor Johnson Tracy, "The Fortune Directory," Fortune (July 1974): 114-15.

29. In interviews with bankers, officials spoke freely about the financial problems but requested that neither their names nor their banks be identified. Many were engaged in East-West negotiations and did not wish to jeopardize their business prospects.

30. Mullin, interview, August 1974.

31. Ibid.

32. U.S. Congress, Senate, Committee on Foreign Relations, Western Investment in Communist Economies, prepared for the Subcommittee on Multinational Corporations by John P. Hardt, George D. Holliday, and Young C. Kim (Washington, D.C.: Government Printing Office, 1974), pp. 39-47.

33. Dan Walsh, correspondence, June 12, 1974.

34. Jack H. Ray, correspondence and interviews, September 1973 to June 1974.

35. U.S. Code, Vol. 4, Sec. 955 (1970); and United States Code Annotated, Title 18, Crimes and Criminal Procedure, Secs. 471 to 1080, Cumulative Annual Pocket Part (St. Paul, Minn.: West Publishing Co., 1973), p. 223.

36. U.S. Code, Vol. 4, Sec. 955 (1970).

37. Ramsey Clark, attorney general, to the secretary of state, May 9, 1967, pp. 3-4

38. Ibid., p. 4.

39. Ibid., p. 5.

40. Ibid., p. 2.

41. Ibid., pp. 7-8; and United States Code Annotated, Title 18, p. 223.

42. Attorney general to secretary of state, May 9, 1967, p. 8.

43. Interview with Jean R. Tartter, deputy director of East-West trade, Department of State, August 1974.

44. U.S. Code, Vol. 4, Sec. 955 (1970).

45. U.S. Department of Commerce, U.S.-Soviet Commercial Agreements, 1972 (Washington, D.C.: Government Printing Office, 1973), pp. 103-07.

46. Ibid.

The purpose of this study was an attempt to analyze the problems and evaluate the feasibility of using a joint venture arrangement to finance U.S. investments in the Soviet Union as embodied in the proposed North Star project. A review of the literature showed that there is a dearth of information concerning the use of the joint venture as a means of financing U.S. investments in the Soviet Union. The information that was available was integrated into a brief background and history of recent joint ventures.

Of critical importance to the North Star project are U.S.-Soviet commercial relations which, during the present period of detente, are difficult to understand because of the proliferation of confusing and conflicting information. In the study, data pertaining to these relations were gathered and clarified through standard research methods. A large part of the research methodology consisted of interviews with representatives of the major U.S. organizations involved in the project.

The primary research question was: ''Is the proposed joint venture arrangement a feasible model for U.S. participation in the North Star project?''

To provide an answer to this question, several subsidiary questions were posed. Answers to the subsidiary questions are presented first, as follows.

1. What arrangements are used in other parts of the world to finance exploration and trade in natural gas?

Foreign direct investment practice has evolved from an insistence on majority equity in the period prior to World War II to a grudging acceptance of joint venture arrangements with minority equity, or even service contract arrangements without equity. All three arrangements coexist in the business world today, and joint ventures are operating in both the planned and the free-market economies.

During the decades of 1920 and 1930, Western business operated in the Soviet Union on the basis of concessions. Political factors played a role in the concessions and eventually caused them to be phased out. Political factors are present in today's environment, but the political environment, both within the Soviet Union and worldwide, has changed since that earlier period.

Control is an important function in any business arrangement, and the efficient operation of an undertaking requires adequate provision for the exercise of effective control. The emphasis on nationalism and control over natural resources by local nationals in the post-World War II period increased the political and economic risks of foreign investment. The joint venture arrangement evolved as an acceptable means of conducting business.

One of the early contractual joint venture agreements was signed on March 3, 1969, by the National Iranian Oil Company (NIOC) and five West European firms for the exploration and development of oil production in South Iran. The foreign firm possessed no equity rights, but agreed to undertake oil exploration and production operations in return for guaranteed rights to purchase oil, if found, at favorable prices. The rights and obligations of all parties were set forth in a contractual agreement.

In the North Star proposal, it is planned that performance obligations will be set forth in firm legal contracts. Through these contracts, the Soviets will be required to deliver natural gas, and the United States will be required to buy it. Financing of the project will be based on the subject contracts which, in turn, will function as collateral instruments for the granting of loans. Failure to deliver gas as specified in the contract would place the Soviets in default on their loans, with consequent detrimental effect on their business reputation and credit rating. By June 1974, all substantive issues were agreed upon, except price, between American and Russian negotiators.

2. Does the U.S. natural gas shortage justify participation in North Star, and would the Soviet gas be cost competitive in the U.S. market?

A natural gas supply shortage has existed in the United States since 1968, when the yearly discovery of new gas reserves fell below annual consumption. The first curtailment of deliveries occurred in 1969, and the problem has become progressively worse. Throughout the United States, the prevalent trend is curtailment of new subscribers and reduction of deliveries to current customers. Official government spokesmen forecast the gas shortage to continue for 20 years.

The cost of generating gas from alternative sources, coal and petroleum distillates, is extremely expensive. Difficulties are compounded by lack of technology and shortage of raw materials. These limitations preclude resolution of the gas shortage from domestic resources on the basis of known technology. If domestic ongoing research and development succeeds, significant quantities of gas would flow by 1985 or 1990, but not in sufficient quantity to satisfy demand. Importation of natural gas under the North Star proposal would be cost competitive with gas produced domestically from alternate sources. A decision in 1975 could cause North Star gas to flow into the American supply system in 1982.

A series of advantages would ensue from North Star. Gas supplies of the United States would be augmented by the import of 18.3 trillion cubic feet (TCF) of gas. Adverse impact on the U.S. balance of payments would be eased because the energy would be obtained through barter trade. Soviet dollar earnings under the proposal would be tied to the purchase of American goods, thus promoting Soviet-American trade and creating, potentially, more than 250,000 man-years of jobs for the U.S. economy.

Analysis of the proposed source of North Star gas, the Urengoi gas field in North Central Siberia, has revealed it to be the largest known gas field in the world. The North Star proposal would require the Soviets to commit approximately 20 percent of the proven capacity of the Urengoi field to meet the delivery requirement to the United States of 2.1 billion cubic feet per day (BCFD) for a 25-year period. The construction of a pipeline and a liquefaction facility in the Soviet Union would require U.S. financing of exports in the amount of $3.7 billion. Ships carrying liquefied natural gas (LNG) and a receiving terminal in the United States would require an additional expenditure of $3 billion.

3. Are Soviet natural gas reserves adequate to support North Star?

4. Do Soviet gas production and consumption trends permit the allocation of sufficient gas to North Star to assure deliveries on a reliable basis?

Russia possesses the largest reserves of natural gas in the world. The reserves are adequate to support domestic requirements and to provide a surplus for export. The bulk of the reserves, however, is located in Siberia and in Central Asia. The rugged terrain and harsh climate have made extraction of the gas difficult and expensive. The Soviets have succeeded in developing the gas resources, but progress is slow and the results obtained fall short of established goals. Access to Western technology and management would reduce the cost and improve the efficiency of Soviet operations. World gas technology is well developed and widely distributed, so that the Soviets could obtain the needed technology from the United States, Western Europe, Canada, or Japan.

Projection of the production and consumption trends for natural gas in the Soviet economy indicates that adequate supplies exist to support North Star. Management policies employed in the USSR traditionally overallocate existing supplies. Therefore, all gas produced could be utilized in the Soviet domestic economy. Greater economic returns would accrue to the Soviet economy, however, by trading natural gas for Western technology.

Official Soviet policy pronouncements have supported trade with the West, and particularly with the United States. The legal foundation, in the sense of a formal document approved by the party and the government, has been enacted into law in the Soviet Union. Soviet ideological constraints exist and will continue to be a factor throughout the planning and implementation of North Star. The Soviet Union appears willing, however, to bend ideological constraints to the extent necessary to accommodate the North Star project.

5. Is North Star feasible from the point of view of the Soviet balance of payments and debt service?

Soviet trade with the West was in a deficit position by approximately $250 million during the decade of the 1960s. The trade deficit increased sharply in the 1970s, to reach its highest level, $1,749 billion, in 1973. Soviet exports to Western nations consist primarily of raw materials and energy products. The increase in world prices for primary commodities will aid the Soviets in handling their trade deficit with the West. Also, the increase in the price of gold provides added resources to cover Soviet trade deficits. The increasing Soviet appetite for Western technology, however, and decreasing supplies of petroleum available for export are expected to cause deficit pressures on the Soviet balance of payments for the indefinite future. Until such time as Soviet manufactured goods are able to compete in Western markets, the Russians will have to rely on the export of primary products to cover their hard-currency trade deficits.

The impact of North Star on the Soviet balance of payments would be strongly positive and would aid in solving the persistent deficits in trade with hard-currency countries. This factor constitutes an incentive for the Soviets to sign the North Star agreement and to meet the delivery terms of the agreement after it is signed.

The Soviet debt service burden decreased from 20 to 17 percent during the period from 1972 to 1973, and was due in a large part to the increased prices the Russians received for exports of primary commodities to hard-currency countries. Also, the Soviets sold approximately $1 billion of their gold reserves.

As a general rule, creditors express concern when the debt service ratio of a country exceeds 20 percent, because experience has shown that repayment problems are encountered at this level of debt. The

Soviet Union has the capability of bearing a larger debt ratio because
it possesses a large gold reserve. In addition, part of the Soviet debt
is self-liquidating; that is, it is tied to repayment in export products
and not hard currency. Also, the annual Soviet gold production, if
sold at current prices, would provide additional hard-currency
reserves to meet debt payments. The Soviet record of debt repayment
since World War II and the strong potential for hard-currency earnings
make the Soviet Union a good credit risk.

The North Star project promises extensive benefits for the Soviet
economy in terms of technology transfer and improved management.
The achieving of real benefits, however, will depend on long-range
continuing exchanges with the West. The Soviet political leaders and
managers expect large increases in productivity as a result of technology imports. Because productivity increases depend on many
factors, including technology, motivation, and the managerial system,
failure to achieve the expected results is likely to cause disappoint-
ment. Ideology is not likely to permit the Soviet Union to open its
society to the degree necessary to achieve maximum economic benefits; nevertheless, concessions already granted reflect a willingness
to compromise.

6. Will U.S. law permit export of the technology required?

The U.S. Constitution grants to the Congress the power to regulate
foreign commerce and to the president the power to administer the
regulations enacted. Friction between the two branches of government
has hindered but has not precluded commercial relations with the
Soviet Union.

The Trade Act of 1974 provides policy direction for commercial
relations with the Soviet Union. It permits the granting of most-
favored-nation (MFN) status and trade credits to the Soviet Union
provided the president makes a determination that the USSR is not
restricting the emigration of its citizens or discriminating against
those who desire to emigrate.

The emigration issue is one of ideology which shows a continuing
thread in U.S. history—an idealistic stand taken in support of human
rights—and is one on which the United States has refused to yield in
the past. In 1913, President Taft abrogated a trade agreement and
withdrew MFN status from Russia because Russian passport regulations discriminated against U.S. citizens of Jewish ancestry. The
United States refused to sign a new trade agreement in spite of the
fact that Russia was an ally during World War I.

The new trade legislation also contains procedures limiting the
export of technology. Existing and proposed legislation restricts the
export of technology that could make a significant contribution to the
military potential of a nation and that might later be used to the

detriment of the national security of the United States. In the bill, procedures are established for interagency review prior to the granting of an export license. Since the Soviets have demonstrated the ability to develop and use gas resources from permafrost regions and since the desired technology is available from other nations, it is believed that American export licenses will be granted. There will probably be administrative delays, however, occasioned by review procedures.

7. Do U.S. tariff or dumping regulations affect North Star?

In 1923, the United States adopted a policy of unconditional MFN treatment, under which a nation agrees to extend, automatically, to other nations any trade concession or advantage it grants to a third nation. In 1951, specific legislation precluded the extension of MFN treatment to Communist nations.

Tariffs pose no barrier to the North Star proposal because the import of LNG is listed as a free item in the U.S. tariff regulations. Therefore, there would be no economic penalty for gas imports from a non-MFN status country. Denial of MFN status to the Soviet Union, however, would have adverse psychological implications for Soviet-American commercial relations. Failure to grant MFN status to Russia would probably hinder a favorable decision on North Star by Soviet leaders, but would not preclude Soviet approval of the project.

Dumping is a practice in which a nation sells a product abroad at a price which is lower than the cost of production or lower than the price of the product in domestic markets. Because the Soviet economy is planned, with wages and prices fixed, it is difficult to ascertain real production costs. Furthermore, the pricing arrangement is not yet settled for the North Star project. When the price issue is settled, it will be necessary to submit the proposed price to the Federal Power Commission (FPC) for approval. In the FPC an analysis will be conducted to confirm that the price is cost competitive in the American market before approval is granted for the import of the gas. In addition, regular delivery of 2.1 BCFD of gas for 25 years is unlikely to cause any disruptions in the market. Therefore, dumping is not likely to be a problem under the North Star proposal.

8. What are the obstacles to the financing of North Star in the U.S. capital market?

Under the present plans, the most important source of U.S. funds for the North Star project is the Export-Import Bank, which operates as an independent agency of the U.S. government. The Export-Import Bank Amendments of 1974, however, restricted loans to the Soviet Union to a total of $300 million with a $40-million sublimit for energy projects. Higher limits may be approved if the president determines that such action is in the national interest and Congress approves such a determination by a concurrent resolution. In addition, specific legislation would be required to increase the authorization limit of the bank to provide funds for financing the North Star project.

In the financing of North Star by U.S. commercial banks, the problem of legal lending limits is encountered. A commercial bank is permitted to loan to a single borrower up to 10 percent of its net worth. It would be possible for the commercial banks to circumvent this law by subdividing North Star into separate projects, such as a pipeline project and a liquefaction project. Or it would be possible for them to loan funds through foreign-owned subsidiary banks. Such action, however, is considered a high risk. Prudent judgment dictates that risk be diversified; hence most bankers prefer to limit loans to the Soviet Union to 10 percent of their net worth.

If Export-Import Bank loan or loan guarantees are provided to cover approximately one-half of the North Star financing funds, the remaining funds could be generated through the commercial bank financing arrangements. Failure of the Export-Import Bank to participate in the project would require commercial banks to exceed current practices concerning prudent risk. The alternative to Export-Import Bank participation would be a consortium of foreign banks. With foreign financing, a requirement to buy foreign equipment would be imposed, which would be detrimental to the U.S. balance of payments.

Another factor involved in the financing of the North Star project is the Johnson Debt Default Act. Foreign governments that are in default to the United States on prior obligations are not permitted to raise capital in the U.S. domestic capital market. Historical practice and formal opinions by three attorneys general of the United States indicate, however, that the act was intended mainly to prohibit the sale of securities in the American capital market. Since credits and loans granted for the purpose of stimulating exports have been ruled legal under the law, and in view of precedent and the fact that North Star is a proposal for long-term barter trade, the project appears to be legal under the Johnson Act.

The Soviet Lend-Lease debt from World War II has been an additional obstacle to improved Soviet-American commercial relations since World War II. The Lend-Lease Settlement Agreement between the United States and the Soviet Union was finally signed, however, on October 18, 1972, with the Soviets agreeing to repay a total of $722 million in cash and annual payments through the period ending July 1, 2001. Three installments would be paid through July 1, 1975, at which time payments would be suspended until such time as the United States granted MFN status to the Soviet Union. Should the United States choose not to grant MFN status to the Soviet Union, the Russians, technically, would not be in default on World War II debts. The restrictions in the Trade Act of 1974 regarding Soviet emigration policy probably will result in denial of MFN status to the USSR.

9. What alternative sources are available to provide financing for North Star?

There is financial risk in any foreign investment, and the granting of equity rights has compensated investors under previous practices of international business. The problem involves security of investment, or the recovery of invested funds, and limitation or diversification of risk. In the North Star project, American goods would be delivered to the Soviet Union in order to develop Soviet natural gas resources, and repayment would be made after the gas field is in production.

There is concern among bankers and involved corporations that the Soviets may renege on deliveries as specified in the contractual agreement. Of course, there are safeguards inherent in the project itself to protect against this danger. Each LNG project is "tailor made" in that it requires special gathering, transportation, and receiving facilities, making it very difficult to divert gas from its planned use. For the North Star project, the lack of LNG ships and the absence of reception terminals in foreign countries would make the export of LNG by the Soviets difficult. The geographic location of North Star facilities would require construction of extensive facilities to permit diversion of the gas to domestic use or diversion into a pipeline for export to European markets.

In addition, failure of the Russians to meet delivery requirements would result in default of loans, causing damage to their business reputation and loss of international credit standing. Such a development would make it difficult for the Soviets to import Western technology over the entire spectrum of their needs and would place the goals established under their annual and Five-Year Plans in jeopardy. The Soviet action of continuing to meet commercial contracts with Western nations, including the United States during the American mining of Haiphong and during the Arab-Israeli war, reflects the importance the Soviets attach to maintaining their international business reputation.

If enabling legislation enacted by the Congress fails to permit participation in the North Star project by the Export-Import Bank, it is unlikely that the project can be financed in the U.S. domestic capital market. An alternative method of financing exists through the formation of an international consortium of banks. There is precedent for such an arrangement in the financing of Algerian gas exports and the proposed financing for the Yakutsk gas project in Eastern Siberia by the Japanese.

The financing of North Star through a consortium of international banks has been investigated by American negotiators for North Star

and has been determined to be feasible. The disadvantages flow from
the requirement that foreign equipment be purchased with the foreign
financing provided, which would cause a delay of approximately 18
months in implementing the project and would result in an adverse
impact on the U.S. balance of payments. Import of gas produced under
consortium financing arrangements would require payment in cash
for that portion of the gas financed by foreign funds. The advantage of
consortium financing arrangements would be diversification of
economic risk and reduction of political risk. United States exposure
would be lower in that less financing would be required, and the Soviet
Union would be even more reluctant to renege on deliveries because
such action would result in default on loans to the United States and
all participating countries.

Thus, on the basis of an analysis of the answers to the subsidiary
questions, it is possible to answer the primary question, "Is the
proposed joint venture arrangement a feasible model for U.S. participation in the North Star project?"

A joint venture is a feasible arrangement for the financing of North
Star. The project will not be easy to implement, for certain compromises will be required by the Soviet Union and the United States, and
certain problems must be resolved.

The first problem area concerns the barter trade concept. This
concept, which forms the theoretical foundation for the North Star
proposal, is compatible with trade theory in that a comparative
advantage exists. Economic benefits would accrue to both the United
States and the Soviet Union, but it is very difficult to determine precisely the value of the benefits. In commodity trade, precise values
of goods are known, and the terms of trade allocate the gains from
trade to each nation. In North Star, the commodity natural gas would
be exchanged for equipment, technology, and managerial know-how
under special financial arrangements. Some aspects of the exchange
are subjective and not amenable to precise values. In addition, worldwide inflation requires a pricing adjustment which spans a period of
25 years. The pricing arrangement that would satisfy the above
uncertainties has not yet been found. Uncertainties demand that a
flexible pricing arrangement be reached which would compensate for
future contingencies. Soviet managers generally dislike a flexible
pricing arrangement because it causes disruptions in the functioning
of their planned economy. American managers would have difficulty
operating under a flexible pricing mechanism because the FPC —
approves the importation of gas on the basis of price. Also, local
utility commissions fix the rates for retail sale of gas, making
adjustments very difficult in the American market.

The second major problem area involves the restrictions on
granting MFN status and extension of credits to the Soviet Union

contained in the Trade Act of 1974 and in the Export-Import Bank
Amendments of 1974. The credit limitation to a total of $300 million
with a $40-million sublimit for energy projects effectively precludes
the use of credits to promote expansion of U.S.-Soviet trade. Provisions in the law do permit higher limits, however, if there is a
presidential determination of national interest approved by a concurrent resolution of Congress.

In order to permit efficient operation of the North Star project
and to achieve the desired economic benefits, certain compromises
are necessary in the operation of the Soviet planned economy.
Reasonably free access to the Russian areas involved must be provided for U.S. managers, and an opening up of Soviet society is
necessary to achieve the full economic benefit of gas technology transfer. To the extent that controls are maintained on either production
or operation of the project, or on the movement of personnel, the
returns from the project will be less than optimal. The Soviet Union
has demonstrated a willingness to compromise to the extent necessary
to gain approval of desired projects, but not to the extent necessary to
achieve maximum efficiency. Therefore, returns from North Star are
expected to be less than optimal.

Assuming the president determines that North Star is in the national
interest and Congress approves the determination by a concurrent
resolution, specific authorizing legislation will be required to provide
the funds needed for North Star. The loan demands are large, and
Export-Import Bank authorized funds are limited. Therefore, special
authorization is needed. Denial of this authorization would preclude
the financing of North Star in the U.S. domestic capital market.

An alternative to Export-Import Bank participation would be a
mixed international consortium of banks to provide the required
financing. The arrangement is feasible, but would have a detrimental
impact on the U.S. balance of payments and would delay implementation of North Star by approximately 18 months. An advantage of the
international consortium financing arrangement would be to minimize
the political risk and to diversify the economic risk of the project.
United States financial exposure would be reduced and foreign nations
would be participants in the projects, thus placing additional pressure
on the Soviets to meet all terms of the North Star contract.

The conclusion of the study is that the joint venture is a feasible
arrangement for U.S. participation in the North Star project. The
investment is sound and would provide economic benefits for both
nations. Political and ideological factors constitute the principal
barriers to successful implementation of the project; nevertheless,
the project is feasible even within the existing institutional structures.
Perhaps an entirely new approach, such as an International East-West

Trade Bank, independent of both U.S. and Soviet political and ideological influences, could be created to implement large projects
similar to North Star. Regardless of the type of arrangement
employed to manage the finances, it is highly likely that the Soviet
type of joint venture arrangement described in this study will become
more common in international business transactions of the future.

APPENDIX

TABLE A.1

U.S. Trade with USSR, 1972, 1973, and 1974
(Thousands of dollars)

Commodity	1972	1973	1974
EXPORTS, TOTAL	542,214	1,194,651	609,248
Food, Beverages and Tobacco	365,767	842,226	287,740
Wheat	154,834	555,613	124,130
Corn	30,762	858	—
Rye	—	41,957	—
Oats	12,102	—	—
Grain sorghums	3	303	518
Citrus fruit, fresh	—	1,096	947
Nuts, edible, except oil nuts	1,063	1,672	4,334
Hops	—	1,941	3.073
Sugars, sirups, and molasses, except cane and beet sugar	—	325	112
Soybean oilcake and meal	—	—	493
Cigarettes	522	387	713
Crude Materials	71,456	77,568	24,913
Cattle hides, undressed	9,557	1,108	7,877
Calf and kip skins, undressed	744	—	31
Sheep and lamb skins, undressed	—	—	4,519
Soybeans	52,561	71,959	—
Cottonseed	—	115	—
Rubber, synthetic	—	—	3,494
Woodpulp, chemical, dissolving grades	8,027	3,328	5,152
Manmade fibers, noncellulosic	496	749	2,973
Natural abrasives, except industrial diamonds	—	200	—
Clay and other refractory minerals	43	67	310
Crude materials, other	29	44	557
Mineral Fuels and Related Materials	—	26	1,336
Lubricating oils and greases	—	1	702
Pitch, petroleum coke, and bitumens	—	19	632
Oils, fats, and waxes	1,701	5,586	—
Linseed oil	1,700	5,586	—

(Continued)

179

Commodity	1972	1973	1974
Chemicals	20,976	16,785	8,017
Coal-tar and other cyclic inter-mediates, except benzene	288	344	964
Rubber compounding chemicals	5	218	217
Plasticizers	306	—	—
Pesticides and synthetic organic agricultural chemicals	154	1,092	702
Alcohols and polyhydric alcohols	376	77	1,293
Organic chemicals, other	1,621	2,570	5,407
Halogen and sulfur compounds of nonmetals or metalloids	603	109	235
Metallic oxides, pigment grade	172	—	659
Oxides and hydroxides of strontium, barium, or magnesium	961	880	617
Aluminum oxide	12,835	2,800	473
Chemical elements and inorganic oxides and halogen salts, other	158	133	3,465
Aluminum sulfate and other aluminum compounds	—	1,418	55
Sodium and potassium compounds	526	319	—
Pigments, paints, and related materials	262	524	659
Glycosides, glands, and vaccines in bulk	104	517	495
Vitamins and fish liver oils, for retail	—	84	2
Medicinal preparations, other than vitamins for retail	399	425	523
Polymerization plastic materials	269	1,518	5,027
Vulcanized fiber and cellulosic plastic materials	341	1,805	2,344
Plastic materials, other	14	109	218
Insecticides, fungicides, herbi-cides, and disinfectants	1,275	591	1,705
Antiknock preparations and other prepared additives	—	11	2,519
Catalysts, compound	—	155	10
Chemicals, other	308	1,086	428

Commodity	1972	1973	1974
Manufactures classified chiefly by material	10,253	34,653	27,429
Leather, cattle hide and kip side	1,388	3,606	1,445
Leather, other	163	193	—
Rubber materials	310	719	996
Rubber manufactures, finished	1,574	4,183	7,124
Paper, paperboard, and manufactures	40	78	150
Yarn and thread of manmade fibers, other than rayon or acetate spun yarn	4,268	8,848	5,804
Coated or impregnated textile fabrics and products	667	1,704	10
Refractory brick and refractory construction materials	584	60	1,669
Refractory products, except construction materials	487	—	4
Iron or steel plates and sheets, uncoated	—	5,066	4,950
Iron or steel plates and sheets, coated, except tin plate	—	146	1,568
Oil pipe of iron or steel	—	5,089	1,260
Iron or steel pipes and tubes, other than cast iron pipe or oil pipe	—	3,546	2
Iron and steel structures and finished parts	—	372	—
Metal containers for storage or manufacturing use	—	123	—
Handtools and tools for machines	326	685	1,592
Metal manufactures, other	63	37	423
Manufactures classified chiefly by material, other	382	196	433
Nonelectric machinery	53,481	181,853	188,189
Internal combustion engines, not for aircraft	36	136	399
Gas turbines, not for aircraft	3	618	5,023
Harvesting machines	190	251	243

(Continued)

Commodity	1972	1973	1974
Wheel tractors	142	194	162
Tracklaying tractors	1,276	26,570	2,787
Agricultural machinery and appliances, other	69	755	254
Electronic computers, including process control computers	3,420	1,919	—
Statistical machines used with punched cards or tape	35	366	2,829
Parts for electronic data processing machines	655	1,577	655
Parts for office machines other than electronic computers	40	137	313
Office machines, other	32	28	290
Drilling machines, metalworking	70	140	—
Gear cutting machines, metalworking	13,202	3,990	1,348
Grinding and polishing machines, metalworking	513	6,758	10,592
Lathes, metalworking, except vertical turret lathes	—	78	972
Milling machines, metalworking	—	567	1,292
Punching and shearing machines, metalworking	7	—	520
Bending and forming machines, metalworking	301	—	1,067
Presses, metalworking	788	18	18,237
Converters, molds, and casting machines, metalworking	217	130	3,774
Rolling mills and parts, metalworking	5	1,896	—
Metalworking machinery, other	5,066	18,957	35,722
Yarn preparation, weaving, and knitting machines	—	961	3,669
Felt making and finishing machinery and parts	—	—	1,164
Auxiliary machines and parts for textile machinery	304	152	870
Textile and leather machinery, other	117	42	336
Printing machinery	313	119	20

Commodity	1972	1973	1974
Food processing machinery, except for grain milling	22	112	104
Construction, excavating, and maintenance machinery, other than road rollers, boring, or crane-type machines	1,792	1,049	803
Boring, mining, and well-drilling machinery	2,071	470	567
Mineral crushing, sorting, mixing, and similar machines	150	622	17,544
Glassworking machinery	695	32	170
Gas generators	—	—	308
Metal processing furnaces and ovens	1	2,975	5,431
Refrigerators and refrigerating equipment, except domestic	43	207	428
Machines for treating nonfood materials with heat or cold	77	444	160
Pumps for liquids	176	18,644	7,436
Air and gas compressors, except refrigeration type	4,187	6,785	27,733
Compressors, refrigeration and airconditioning type	274	80	5
Filtering, purifying, and separating machinery	515	672	2,050
Oil and gas field lifting and loading equipment	1,222	7,770	11
Lifting and loading equipment, other than oil and gas field equipment or underground loaders for mining	3,799	57,880	18,332
Machine tools, parts, and accessories, mineral working	203	17	616
Machine tool parts and accessories, metalworking	1,269	1,466	2,641
Bottling, packaging, and wrapping machinery	55	366	26
Sprayers and spraying equipment	457	654	2,254
Ball and roller bearings	82	223	523

(Continued)

Commodity	1972	1973	1974
Plastics working machinery	2	427	143
Rubber extruding, tire, and rubber processing machinery	2,532	10,538	135
Metal treating and metal powder molding machines	—	220	205
Taps, valves, and similar appliances	235	221	44
Nonelectric machinery, other	6,710	3,620	7,982
Electric machinery and apparatus	7,228	14,474	27,573
Electric power machinery	338	80	2,956
Electric circuit apparatus	88	251	272
Insulated wire and cable	502	532	1,000
Radio and television broadcast equipment	20	113	95
Intercommunication equipment except telephone or telegraph	—	1	1,039
Electronic navigational aids	150	326	59
Telecommunications apparatus, other	98	65	573
Electromedical apparatus, except X-ray apparatus	187	563	435
X-ray and radiological apparatus	102	41	618
Waveform measuring or analyzing instruments	121	141	269
Instruments for measuring or testing electric or electronic characteristics other than waveform or frequency	113	300	132
Nuclear radiation detecting and measuring instruments	211	203	537
Geophysical and mineral prospecting instruments	10	340	264
Physical properties analysis and testing instruments	2,534	4,281	3,687
Electric instruments for measuring or controlling nonelectric quantities, other	597	841	1,893
Electric machinery and apparatus, other	2,158	6,396	13,744

Commodity	1972	1973	1974
Transport equipment	1,353	7,980	9,228
Trucks	—	586	465
Special purpose vehicles, except power cranes and other excavating or drilling type equipment	233	380	492
Parts and accessories for tractors	485	5,702	6,778
Parts and accessories for motor vehicles, except tractors	554	723	747
Truck trailers	4	277	150
Vehicles, except road motor vehicles, other	—	262	2
Aircraft, heavier than air	—	—	439
Miscellaneous manufactured articles	9,163	9,132	12,529
Furniture	11	81	40
Photographic and motion-picture equipment, except cameras, sound equipment, and motion-picture projectors	386	375	170
Medical instruments	256	288	306
Navigational and surveying instruments	17	89	22
Technical models for demonstration	7	352	217
Instruments for measuring variables of liquids or gases	241	6	467
Instruments for physical or chemical analysis	538	953	1,347
Parts for scientific, measuring, or controlling instruments	415	351	672
Scientific, measuring, and controlling instruments, other	378	195	221
Instruments, photographic goods, watches and clocks, other	61	82	230
Recording, dictating, and transcribing machines and parts	171	188	498
Phonographs, videotape recorders, parts, and accessories	24	—	1,454

(Continued)

Commodity	1972	1973	1974
Advertising matter, catalogs, and business publications	26	142	102
Printed matter, other	53	113	326
Plastic packaging and shipping containers, except bags	3,729	2,659	1,781
Articles of artificial plastics, other	2,496	2,969	3,541
Works of art and collectors' items	36	11	737
Other domestic exports	706[d]	1,698	903
Reexports[a]	130	2,671	1,392
IMPORTS, TOTAL	95,536	220,072	350,223
Food, beverages, and tobacco	713	868	1,177
Lobsters	193	—	—
Fish in airtight containers and fish preparations	318	253	118
Molasses, inedible	—	—	468
Food, other	24	66	49
Alcoholic beverages	174	518	538
Crude materials	17,963	11,646	19,195
Sheep and lamb skins, undressed, without wool	—	—	183
Persian lamb and caracul fur, undressed	648	805	360
Rabbit fur, undressed	1	—	176
Sable fur, undressed	1,501	2,116	3,559
Furskins, undressed, other	857	214	444
Industrial diamonds	184	1,090	430
Natural abrasives, other	—	—	317
Clay and other refractory minerals	3	27	572
Asbestos	—	—	123
Iron ore and concentrates	—	—	1,622
Chrome ore	14,056	6,431	9,438
Ash and residues bearing non-ferrous metals	—	—	1,125

Commodity	1972	1973	1974
Bones, ivory, horns, and similar products	222	122	68
Bristles	240	516	406
Licorice root	—	112	—
Plants used in perfumery, pharmacy and insecticides, other	151	75	209
Crude materials, other	100	136	160
Mineral fuels and related materials	7,464	76,524	105,814
Coke, suitable for fuel	—	—	2,379
Petroleum, crude and partly refined	1,952	17,129	24,941
Gasoline and motor fuels, except jet fuel	—	3,760	19,470
Jet fuel and kerosene	—	4,909	14,446
Distillate fuel oils	—	37,778	32,592
Residual fuel oils	5,510	12,943	7,344
Pitch, asphalt, and other bitumens	—	—	4,585
Oils, fats, and waxes	1	12	2
Chemicals	1,250	2,307	12,449
Toluene	122	—	—
Organic chemicals suitable for medicinal use	—	1	984
Organic chemicals, other	—	35	3,489
Metallic oxides, pigment grade	—	14	1,121
Barium dioxide, hydroxide and oxide	—	49	541
Antimony oxide	—	61	50
Chrome green and other chromium oxides and hydroxides	260	—	—
Chemical elements and inorganic oxides and halogen salts, other	62	545	889
Sodium chromate and dichromate	507	—	69
Inorganic chemicals, except elements, oxides, and halogen salts, other	156	875	1,011
Isotopes, radioactive elements, and compounds	1	173	190

(Continued)

Commodity	1972	1973	1974
Essential oils, perfume, and flavor materials	108	40	64
Potassic fertilizers and materials	—	—	1,459
Cartridges and shells for small arms	2	3	97
Casein	—	217	2,039
Gelatin, inedible, and animal glue	19	280	346
Chemicals, other	13	15	99
Manufactures classified chiefly by material	63,666	122,715	204,686
Plywood, including wood veneer panels	641	1,203	1,267
Wood manufactures, except furniture	31	124	116
Boxes of paper, paperboard, or papier mache	26	56	77
Cotton fabrics, woven	527	2,086	953
Wadding, wicks, and textile fabrics for industrial use	—	—	87
Carpets, carpeting, and rugs	86	170	59
Drawn or blown glass, unworked	2,130	3.251	1,675
Diamonds, except industrial, not set or strung	13,435	17,260	11,875
Precious and semiprecious stones, and pearls, other	52	19	39
Nonmetallic mineral manufactures and gems, other	24	66	38
Ferrosilicon	—	60	706
Pig iron and ferroalloys, other	—	30	684
Platinum	18,540	13,710	19,659
Iridium	135	—	313
Palladium	19,511	47,440	75,903
Rhodium	1,251	4,887	18,093
Silver, platinum, and platinum group metals, other	5,272	14,304	20,215
Copper and copper alloys, unwrought	—	272	1,835
Nickel and nickel alloys, unwrought	246	10,538	39,939

Commodity	1972	1973	1974
Aluminum, unwrought, not alloyed	56	—	—
Zinc, unwrought, not alloyed	—	2,777	261
Titanium	748	3,850	10,257
Nonferrous metals, other	839	484	444
Chains and parts of iron or steel	54	64	1
Metal manufactures, other	13	26	107
Manufactures classified chiefly by material, other	51	37	82
Nonelectric machinery	49	42	1,713
Wheel tractors, except garden, suitable for agriculture	—	—	325
Tractors, except wheel tractors for agriculture	—	—	885
Boring machines and vertical turret lathes, metalworking	—	—	75
Lathes, metal cutting, except vertical turret lathes	—	—	117
Nonelectric machinery, other	49	42	311
Electric machinery and apparatus	396	76	66
Electric power machinery and switchgear	386	1	11
Electron tubes and parts	7	47	—
Electric machinery and apparatus, other	3	27	55
Transport equipment	16	39	115
Road motor vehicles	3	11	115
Miscellaneous manufactured articles	3,199	4,463	3,861
Footwear	16	128	49
Still cameras, parts, and flash apparatus	283	250	200
Scientific, measuring, and controlling instruments, other	7	46	70
Exposed photographic and motion-picture film	36	35	39
Musical instruments and sound reproducers	54	38	58
Printed matter	78	65	94
Dolls	22	82	25
Nonmilitary firearms, except revolvers and pistols	([b])	10	1,214

(Continued)

Commodity	1972	1973	1974
Toys, games, and sporting			
goods, other	52	19	43
Works of art	31	578	640
Antiques	168	151	246
Stamps	371	453	404
Works of art and collectors'			
items, other	325	190	157
Jewelry and related articles of			
precious metals	1,690	2,283	400
Carved or molded goods	22	50	101
Miscellaneous manufactured			
articles, other	43	85	122
Other imports[c]	819	1,381	1,145
Returned goods	727	1,300	1,040
Live animals not for food	87	3	—

[a]Merchandise of foreign origin which entered the United States as imports and which at the time of export were in substantially the same condition as when imported.

[b]Less than $500.

[c]Includes entries under $250.

[d]Includes relief shipments.

Note: Figures may not add because of rounding.

Source: U.S. Department of Commerce, Bureau of East-West Trade, Export Administration Report, 1st Quarter 1975 (Washington, D.C.: Government Printing Office, 1975).

PUBLIC DOCUMENTS

Commission on International Trade and Investment Policy. United States International Economic Policy in an Interdependent World. 3 vols. Commission's report to the president. Washington, D.C.: Government Printing Office, 1971.

Council on International Economic Policy. International Economic Report of the President. Washington, D.C.: Government Printing Office, February 1974.

Export-Import Bank of the United States. Eximbank and the World of Exports. Statement of condition, fiscal year 1973. Washington, D.C.: Government Printing Office, 1974.

Hungary. Magyar Koz'ony Number 67 [Hungarian Law Number 67]. Budapest, August 7, 1970.

"Joint Adventures." 48 Corpus Juris Secundum 801 (1947).

U.S. Central Intelligence Agency. The Soviet Economy in 1973: Performance Plans and Implications. A(ER) 74-62. Washington, D.C.: Central Intelligence Agency, July 1974.

U.S. Code, vol. 4 (1970); Supp. 2, Title 50, App. 2402 (1970).

U.S. Congress. House. Export-Import Bank Amendments of 1974. P.L. 93-646; 88 Stat. 2333, 93rd Cong., 2nd sess., 1975, H.R. 15977.

_____. Trade Act of 1974. P.L. 93-618, 93rd Cong., 2nd sess., 1975, H.R. 10710.

U.S. Congress. House. Committee on Banking and Currency. International Economic Policy. Hearings before the Subcommittee on International Trade on H.R. 774, H.R. 13838, H.R. 13839, H.R. 13840, 93rd Cong., 2nd sess., 1974.

U.S. Congress. House. Committee on Foreign Affairs. U.S.-Soviet Commercial Relations: The Interplay of Economics, Technology

Transfer, and Diplomacy. Prepared for the Subcommittee on National Security Policy and Scientific Developments by John P. Hardt and George D. Holliday. Washington, D.C.: Government Printing Office, 1973.

U.S. Congress. House. Committee on Science and Astronautics. Energy Research and Development—An Overview of Our National Effort. Hearings before the Subcommittee on Energy, 93rd Cong., 1st sess., 1973.

———. The Technology Balance: U.S.-USSR Advanced Technology Transfer. Hearings before the Subcommittee on International Cooperation in Science and Space, 93rd Cong., 1st and 2nd sess., 1973.

U.S. Congress. Joint Economic Committee. New Directions in the Soviet Economy. Pt. II-A: Economic Performance, 89th Cong., 2nd sess., 1966.

———. Soviet Economic Outlook. Hearings, 93rd Cong., 1st sess., July 19, 1973.

———. Soviet Economic Prospects for the Seventies: A Compendium of Papers Submitted to the Joint Economic Committee. Edited by John P. Hardt. Joint Committee Print. Washington, D.C.: Government Printing Office, 1973.

U.S. Congress. Joint Economic Committee. Subcommittee on Economic Statistics. Comparisons of the U.S. and Soviet Economies, pt. 2, 86th Cong., 1st sess., 1959.

U.S. Congress. Joint Economic Committee. Subcommittee on Priorities and Economy in Government. Allocation of Resources in the Soviet Union and China, Hearings, 93rd Cong., 2nd sess., April 12, 1974.

U.S. Congress. Senate. Committee on Banking and Currency. Export Expansion and Regulation. Hearings before the Subcommittee on International Finance on S. 813 and S. 1940, 91st Cong., 1st sess., 1969.

U.S. Congress. Senate. Committee on Banking, Housing, and Urban Affairs. Export-Import Bank of the United States. Hearings before the Subcommittee on International Finance on S. 1890, 93rd Cong., 1st sess., 1973.

U.S. Congress. Senate. Committee on Banking, Housing, and Urban
    Affairs. The Role of the Export-Import Bank and Export Controls
    in U.S. International Policy. Hearings before the Subcommittee
    on International Finance on S. 1890 and S. 3282, 93rd Cong., 2nd
    sess., 1974.

U.S. Congress. Senate. Committee on Finance. The Trade Reform
    Act of 1973. Hearings before the Committee on Finance on H.R.
    10710, 93rd Cong., 2nd sess., 1974.

U.S. Congress. Senate. Committee on Foreign Relations. Foreign
    Assistance Act of 1971, 92nd Cong., 1st sess., 1971.

_____. Multinational Corporations and United States Foreign Policy.
    Hearings before the Subcommittee on Multinational Corporations
    on Multinational Petroleum Companies and Foreign Policy, 93rd
    Cong., 2nd sess., January 30, 1974.

_____. Western Investment in Communist Economies. Prepared for
    the Subcommittee on Multinational Corporations by John P. Hardt,
    George D. Holliday, and Young C. Kim. Washington, D.C.: Govern-
    ment Printing Office, 1974.

U.S. Congressional Record, 93rd Cong., 1st sess., February 7, 1973,
    vol. 119, pp. 9106-08.

U.S. Department of Commerce. Export Control. 96th quarterly report,
    2nd quarter 1971. Washington, D.C.: Government Printing Office,
    August 16, 1971.

_____. U.S. Commercial Relationship in a New Era. Prepared by
    Peter G. Peterson. Washington, D.C.: Government Printing
    Office, August 1972.

U.S. Department of Commerce. Domestic and International Business
    Administration. Bureau of East-West Trade. East-West Trade.
    Export administration report for fourth quarter, 1973. Washing-
    ton, D.C.: Government Printing Office, 1974.

_____. U.S.-Soviet Commercial Agreements 1972: Texts, Summaries
    and Supporting Papers. Washington, D.C.: Government Printing
    Office, 1973.

U.S. Department of the Interior. Bureau of Mines Minerals Yearbook,
    1971. Washington, D.C.: Government Printing Office, 1972.

_____. Bureau of Mines Minerals Yearbook, 1972. Washington, D.C.: Government Printing Office, 1973.

U.S. Department of State. American Embassy Helsinki Message 0071. January 11, 1974.

U.S. Federal Power Commission. "Future Gas Supplies from Alternate Sources." Preliminary draft of Chap. 10, National Gas Survey, 1974.

_____. National Gas Supply and Demand 1971-90. Staff Report No. 2. Washington, D.C.: Bureau of Natural Gas, Federal Power Commission, February 1972.

_____. Natural Gas Reserves Study. National Gas Survey Staff Report. Revised. Washington, D.C.: Federal Power Commission, September 1973.

U.S. Statutes at Large, vol. 83 (1969); vol. 86 (1972).

U.S. Tariff Commission. Impact of Granting Most-Favored-Nation Treatment to the Countries of Eastern Europe and the People's Republic of China. Prepared by John E. Jelacic. Staff Research Studies No. 6. Washington, D.C.: U.S. Tariff Commission, 1974.

_____. Pig Iron from East Germany, Czechoslovakia, Romania, and the USSR. TC Pubn. 265. Washington, D.C.: U.S. Tariff Commission, September 1968.

_____. Tariff Schedules of the United States Annotated (1972). TC Pubn. 452. Washington, D.C.: Government Printing Office, 1972.

_____. Titanium Sponge from the USSR. TC Pubn. 255. Washington, D.C.: U.S. Tariff Commission, July 1968.

_____. United States East European Trade Considerations Involved in Granting Most-Favored-Nation Treatment to the Countries of Eastern Europe. Prepared by Anton F. Malish, Jr. Staff Research Studies No. 4. Washington, D.C.: U.S. Tariff Commission, 1972.

USSR. Central Statistical Directorate. Narodnoye Khozyajstvo SSSR v. 1970 G [National Economy of the USSR in 1970]. Moscow: Statistika, 1971.

USSR. Central Statistical Directorate. Narodnoye Khozyajstvo SSSR v. 1972 G [National Economy of the USSR in 1972]. Moscow: Statistika, 1973.

USSR. Ministry of Foreign Trade. Vneshnyaya Torgovliya [Foreign Trade]. Soviet yearbooks, 1960 through 1973. Moscow: Mezhdunarodnye Otnoshenia, 1960 through 1973.

TESTIMONY

Boretsky, Michael. "Comparative Progress in Technology Productivity and Economic Efficiency: USSR versus U.S.A." Testimony in U.S. Congress, Joint Economic Committee, New Directions in the Soviet Economy. Pt. II-A: Economic Performance, 89th Cong., 2nd sess., 1966.

Bucy, J. Fred, executive vice-president, Texas Instruments, Inc. Testimony in Hearings before the Subcommittee on International Cooperation in Science and Space, Committee on Science and Astronautics, U.S. House of Representatives, The Technology Balance: U.S.-USSR Advanced Technology Transfer, December 5, 1973.

Campbell, Robert W. "Some Issues in Soviet Energy Policy for the Seventies." In U.S. Congress, Joint Economic Committee, Soviet Economic Prospects for the Seventies, ed. John P. Hardt, Joint Committee Print, June 27, 1973.

_____. "Technology Transfer in Expanded Commercial Relations between the U.S. and USSR." Testimony in Hearings before the Subcommittee on International Trade, Banking and Currency Committee, U.S. House of Representatives, International Economic Policy, April 24, 1974.

Carameros, George D., Jr. "The Yakutia Liquefied Natural Gas Project." Prepared statement delivered at Hearings before the Subcommittee on Multinational Corporations, Committee on Foreign Relations, U.S. Senate, June 17, 1974.

Casey, William J. "Statement," in Hearings before Subcommittee on International Trade, Committee on Banking and Currency, U.S. House of Representatives, International Economic Policy, April 1974.

Casey, William J. Testimony in Hearings before Subcommittee on
    International Finance, Committee on Banking, Housing, and Urban
    Affairs, U.S. Senate, The Role of the Export-Import Bank and
    Export Controls in U.S. International Economic Policy, April 1974.

Dent, Frederick B., secretary of commerce. Testimony before the
    Subcommittee on Multinational Corporations, Committee on
    Foreign Relations, U.S. Senate, July 17, 1974.

Farrell, John T. "Soviet Payments Problems in Trade with the West."
    In U.S. Congress, Joint Economic Committee, Soviet Economic
    Prospects for the Seventies, ed. John P. Hardt, Joint Committee
    Print, June 27, 1973.

Hardt, John P. "Summary" of Soviet Economic Prospects for the
    Seventies, ed. John P. Hardt, Joint Committee Print, June 27,
    1973.

Holzman, Franklyn D. "East-West Trade and Investment Policy
    Issues." In Commission on International Trade and Investment
    Policy, United States International Economic Policy in an Inter-
    dependent World, vol. 2, July 1971.

Kearns, Henry. Testimony in Hearings before the Subcommittee on
    International Finance, Committee on Banking, Housing, and Urban
    Affairs, U.S. Senate, Export-Import Bank of the United States,
    1973.

Lazarus, Steven. Testimony in Hearings before the Joint Economic
    Committee, U.S. Congress, Soviet Economic Outlook, July 19,
    1973.

_____. Testimony in Hearings before the Subcommittee on Inter-
    national Cooperation in Science and Space, Committee on Science
    and Astronautics, U.S. House of Representatives, The Technology
    Balance: U.S.-USSR Advanced Technology Transfer, December 5,
    1973.

Lee, J. Richard. "The Soviet Petroleum Industry: Promise and
    Problems." In U.S. Congress, Joint Economic Committee,
    Soviet Economic Prospects for the Seventies, ed. John P. Hardt,
    Joint Committee Print, June 27, 1973.

Ray, Jack H. "The North Star Soviet Liquefied Natural Gas (LNG)
    Import Project." Testimony in Hearings before the Subcommittee

on International Trade, Committee on Banking and Currency, U.S. House of Representatives, International Economic Policy, April 26, 1974.

Shields, Dr. Roger E., deputy assistant secretary of defense, international economic affairs. "National Security Impact of U.S. Capital Investment and High Technology Transfers to the USSR." Testimony in Hearings before the Subcommittee on Multinational Corporations, Committee on Foreign Relations, U.S. Senate, July 18, 1974.

Sutton, Anthony C. Testimony in Hearings before the Subcommittee on International Trade, Banking and Currency Committee, U.S. House of Representatives, April 24, 1974.

Wilson, Edward T., et al. "U.S.-Soviet Commercial Relations." In U.S. Congress, Joint Economic Committee, Soviet Economic Prospects for the Seventies, ed. John P. Hardt, Joint Committee Print, June 27, 1973.

BOOKS

Baibakov, N. K. State Five-Year Plan for the Development of the USSR National Economy for the Period 1971-75. (English translation of the Soviet Five-Year Plan, 1971-75.) Arlington, Va.: Joint Publications Research Service, September 1972.

Beguin, Jean-Pierre. "The Control of Joint Ventures." In Joint International Business Ventures in Developing Countries, pp. 364-418. Edited by Wolfgang G. Friedmann and Jean-Pierre Beguin. New York: Columbia University Press, 1971.

Belorusov, Dmitri V., et al. Osvoyeniye Neftyanykh Mestorozhdenij Zapadnoj Sibiri [Mastering the Petroleum Fields in Western Siberia]. Moscow: Nedra, 1972.

Berliner, Joseph S. Soviet Economic Aid. New York: Frederick A. Praeger, 1958.

Black's Law Dictionary. Rev. 4th ed. (1968).

Buyalov, N. I., et al. Metodike Otsenki Prognonznykh Zapasov Nefti
i Gaza [Methods for Evaluating Predicted Reserves of Gas and
Oil]. Leningrad: Gos Top Tekh Izdat, 1962.

Brzezensi, Zbigniew, and Samuel Huntington. Political Power: U.S.A./
USSR. New York: Viking Press, 1963.

Campbell, Robert W. The Economics of Soviet Oil and Gas. Baltimore:
Johns Hopkins Press, 1968.

Ceausescu, Nicolae. Law on Foreign Trade and Economic and
Technico-Scientific Cooperation Activities in the Socialist
Republic of Romania. Bucharest: Chamber of Commerce of the
Socialist Republic of Romania, 1971.

Clabaugh, Samuel F., and Edwin J. Feulner. Trading with the Communists. Washington, D.C.: Center for Strategic Studies, 1968.

Business International, S.A. Doing Business with the USSR. Geneva,
November 1971.

Condoide, Mikhail V. Russian-American Trade. Columbus: Ohio
State University, 1946.

Franko, Lawrence G. Joint Venture Survival in Multinational
Corporations. New York: Praeger Publishers, 1971.

Friedmann, Wolfgang G., and Jean-Pierre Beguin. Joint International
Business Ventures in Developing Countries. New York: Columbia
University Press, 1971.

Friedmann, Wolfgang G., and Leo Mates. Joint Business Ventures of
Yugoslav Enterprises and Foreign Firms. Belgrade, 1968.

Halbouty, Michel T. Geology of Giant Petroleum Fields. Tulsa:
American Association of Petroleum Geologists, November 1970.

Hardt, John P. "Soviet Economic Development and Policy Alternatives." In The Development of the Soviet Economy, pp. 5-6.
Edited by Vladimir G. Treml. New York: Praeger Publishers,
1968.

Herman, Leon. "The Promise of Economic Self-Sufficiency under
Soviet Socialism." In The Development of the Soviet Economy,
pp. 215-16. Edited by Vladimir G. Treml. New York: Praeger
Publishers, 1968.

Hickman, John H. "A New Frontier Opens as the Iron Curtain Rolls
    Down." Financing East-West Business Transactions. AMA
    Management Bulletin No. 119. New York: American Management
    Association, 1968.

Kindelberger, C. P. International Economics. 3rd ed. Homewood,
    Ill.: Richard D. Irwin, Inc., 1963.

Kretschmar, Robert S., Jr., and Robin Foor. The Potential for Joint
    Ventures in Eastern Europe. New York: Praeger Publishers,
    1972.

Laserson, Max B. The American Impact on Russia: Diplomatic and
    Ideological, 1784-1917. New York: Macmillan Co., 1950.

L'vov, Mikhail S. Resursy Prirodnovo Gaza SSSR [Natural Gas
    Resources of the USSR]. Moscow: Nedra, 1969.

The New Encyclopaedia Britannica. 15th ed., Micropaedia, vol. 3,
    Ready Reference (1973).

Pisar, Samuel. Coexistence and Commerce. New York: McGraw-Hill,
    1970.

Rousseau, Jean Jacques. "The Social Contract." Great Books of the
    Western World. Edited by R. M. Hutchins. Chicago: Encyclo-
    paedia Britannica, Inc., 1952.

Schwartz, Harry. The Soviet Economy. Philadelphia: Lippincott, 1965.

Smith, Len Young, and G. Gale Roberson. Business Law: Uniform
    Commercial Code. 3rd ed. St. Paul, Minn.: West Publishing Co.,
    1971.

Spulber, Nicolas. Soviet Strategy for Economic Growth. Bloomington:
    Indiana University Press, 1964.

Sutton, Anthony C. Western Technology and Soviet Economic Develop-
    ment, 1917 to 1930. Stanford, Calif.: Hoover Institution, 1968.

Turabian, Kate L. A Manual for Writers of Term Papers, Theses,
    and Dissertations. 4th ed. Chicago: University of Chicago Press,
    Phoenix Books, 1973.

United States Code Annotated. Title 12, Banks and Banking, Sections
    1-530. St. Paul, Minn.: West Publishing Co., 1972.

United States Code Annotated. Title 18, Crimes and Criminal Pro-
    cedure, Sections 471 to 1080. Cumulative Annual Pocket Part.
    St. Paul, Minn.: West Publishing Co., 1973.

Zaleski, Eugene. "Planning for Industrial Growth." In Development
    of the Soviet Economy, pp. 55-56. Edited by Vladimir G. Treml.
    New York: Frederick A. Praeger, 1968.

STUDIES AND REPORTS

American Gas Association. Natural Gas Supply Problem: Background
    Report. Arlington, Va.: American Gas Association, 1972.

Business International Corporation. 100 Checklists: Decision-making
    in International Operations. New York: Business International
    Corp., 1970.

Dodge, Norton. "Plan and Economy: Discussion." Analysis of the
    USSR's 24th Party Congress and 9th Five-Year Plan. Mechanics-
    ville, Md.: Cremona Foundation, 1971.

Foreign Broadcast Information Service (FBIS)—E. Europe. November
    3, 1971.

Future Requirements Committee. Future Gas Consumption of the
    United States, vol. 5. Denver: University of Denver Research
    Institute, Future Requirements Agency, November 1973.

Samuel Montagu and Co. Ltd. Annual Bullion Review, 1973. London,
    March 1974.

North Star Geological and Reserves Report. Houston: Tenneco Oil Co.,
    March 31, 1972.

North Star Project. A project of Tenneco Inc., Texas Eastern Trans-
    mission Company, and Brown and Root (Houston, 1972).

North Star Project Feasibility Study. Vol. 1: Text. Houston: Brown
    and Root, Inc., Tenneco Inc., and Texas Eastern Transmission
    Corporation, March 31, 1972.

North Star Project Financing Analysis. Houston: Brown and Root,
    Inc., Tenneco Inc., and Texas Eastern Transmission Corporation,
    June 4, 1973.

Peterson, Peter G. A Foreign Economic Perspective. Washington,
    D.C.: Council on International Economic Policy, December 27,
    1971.

_____. The United States in the Changing World Economy. Washing-
    ton, D.C.: Council on International Economic Policy, December
    27, 1971.

Petroleum Extension Service. Oil Pipeline Construction and Mainte-
    nance. Austin: University of Texas, April 1973.

Strishkov, V. V. "The Minerals Industry of the USSR." Preprint
    from the 1971 Bureau of Mines Minerals Yearbook, U.S. Depart-
    ment of the Interior. Washington, D.C.: Government Printing
    Office, 1971.

Tenneco, 1972 Annual Report. Houston: Tenneco Inc., 1972.

Tenneco, Special International Issue, vol. 7 no. 2. Houston: Tenneco
    Inc., summer 1973.

U.S. Natural Gas Delegation. Gas in the Soviet Union. Arlington, Va.:
    American Gas Association, Inc., 1970.

MICROFORM REPRODUCTIONS

Dissertation Abstracts International. A: The Humanities and Social
    Sciences. Ann Arbor: Xerox University Microfilms, vol. 30, no. 1
    (July 1969) to vol. 34, no. 8 (February 1974).

ARTICLES AND PERIODICALS

Almond, Gabriel A. "Book Reviews." American Political Science
    Review, December 1964, pp. 976-77; and June 1965, pp. 446-47.

American Gas Association, Gas Supply Committee. "Summary of
    United States Natural Gas Statistics for the Period 1945-73."
    Gas Supply Review, supplement, May 15, 1974.

Aug, Stephen M. "Arab Prices a Mystery to Exxon Chief." Washington
    Star-News, January 12, 1974, p. C-back page.

Baibakov, N. K. "O Gosudarstvennom Plane Razvitia Narodnovo
    Khozyaistva SSSR na 1974 God; Vneshniye Ekonomicheskiye
    Svyazi" [Concerning the Government Plan for Development of
    the National Economy of the USSR during 1974]. Pravda,
    December 13, 1973, p. 3.

"Ban on Credit to Red Bloc Ended." Washington Evening Star,
    December 1, 1971, p. 1.

Buzuluk, O. "Sever-1" [North-1]. Krasnaya Zvezda [Red Star],
    July 15, 1973, p. 4.

Dewan, George. "Professors Vindicated by Energy Crisis."
    Washington Post, January 1, 1974, p. D-9.

"Direktivy XXIV S'yezda KPSS po Pyatiletnemu Planu Razvitiya
    Narodnovo Khozyajstva SSSR na 1971-75 Gody" [Direction of
    the 24th Congress of C.P.S.U. Concerning the Five-Year Plan
    for the Development of the National Economy of the USSR during
    the Period 1971-75]. Pravda, February 14, 1971, p. 2.

Ebel, Robert E. "Russia Falls Short of Pipe Line Goals." Pipe Line
    Industry, November 1973, pp. 33-36.

_____. "Russians 'Think Big' in Future Pipe Line Plans." Pipe Line
    Industry, November 1972, p. 30.

Farrer, Dean G. "The Soviet Industrial Purchasing Agent." Journal
    of Purchasing, November 1969.

"The First Contract for Kama River." Business Week, January 1,
    1972.

Franko, Lawrence G. "Joint Venture Divorce in the Multinational
    Company." Columbia Journal of World Business, May-June
    1971, pp. 13-22.

Freeman, S. David. "The Forgotten Energy Agencies." Washington Post, February 10, 1974, p. B-1.

"Gazovaya Promyshlennost" [Gas Industry]. Ekonomicheskaya Gazeta [The Economic Gazette], February 1973, p. 2.

Hale, Dean. "International LNG Movements Continue to Expand." Pipeline and Gas Journal, December 1973, pp. 61-63.

Hammer, Armand. "American Entrepreneur—First Foreign Concessionaire in the Soviet Union." American Review of East-West Trade, March-April 1970, pp. 14-22.

Hardt, John P. "West Siberia: The Quest for Energy." Problems of Communism, May-June 1973, pp. 28-29.

"How's Brezhnev's Credit Rating?" Forbes, June 15, 1973, pp. 33-34.

"Informatsionnoye Soobshcheniye: O Plenume Tsentralnovo Komiteta Kommunisticheskoj Partii Sovetskovo Syuza" [Informational Notes: Concerning the Plenum of the Central Committee of the Communist Party of the Soviet Union]. Pravda, December 12, 1973, p. 1.

"Is There a Ford in Russia's Future?" Business Week, April 18, 1970.

Isaacs, Stephen. "End Seen to Trade Impasse." Washington Post, September 7, 1974, p. 1.

Izvestia, February 3, 1973, p. 1; March 20, 1973, p. 1; March 21, 1973, p. 3.

Kaser, Michael. "Soviet Union." International Currency Review, July-August 1973, p. 92.

Khan, Aman R. "Soviet Gas in the Seventies." Pipeline and Gas Journal, October 1972, p. 25.

Knight, John S. "Oil Companies Helped Create the Energy Crisis." Detroit Free Press, November 18, 1973.

Kovalyev, S. "Suverenitet i Internatsionalnye Obyazannosti Sotsialisticheskikh Stran" [Sovereignty and the International Obligations of Socialist States]. Pravda, September 26, 1968, p. 4.

Kraft, Joseph. "Business with Russia." Washington Post, November 16, 1971.

Leontiev, L. "Myth about 'Rapproachment' of the Two Systems." Ekonomicheskaya Gazeta [The Economic Gazette] , December 1966.

Liberman, Yevsei. "The Soviet Economic Reform." Foreign Affairs 46 (October 1967): 53.

Marder, Murrey, and Marilyn Berger. "U.S.-Soviet Grain Deal: Case History of a Gamble." Washington Post, December 7, 1971.

"Namyecheny Vysokiye Rubyezhi" [High Goals Are Set] . Pravda, January 14, 1973, p. 1.

Neumann, Timothy P. "Joint Ventures in Yugoslavia: 1971 Amendments to Foreign Investment Laws." New York University Journal of International Law and Politics 6 (summer 1973): 296.

Oberdorfer, Don. "Soviets Unblock Japan's Role in Siberian Oil and Gas." Washington Post, March 10, 1974, p. A-20.

Patolichev, N. "Vzaimovygodnoye Sotrudnichestvo" [Mutually Advantageous Cooperation] . Pravda, December 27, 1973, p. 4.

Pearson, John. "The Big Breakthrough in East-West Trade." Business Week, June 19, 1971.

Permikhin, Yu. "A Basic Support, Not Just Help." Pravda, March 3, 1973, p. 2.

Pravda, January 30, 1973; March 5, 1973; December 15, 1973; February 26, 1974.

Rand, Christopher. "The Great Oil Crisis Manipulation." Washington Post, January 20, 1974.

Reistrup, J. V. "Truck Deal to Russia Approved." Washington Post, November 19, 1971.

"Russia's Needs Seen Pushing Its Purchases of U.S. Grain above Levels of Nixon Pact." Wall Street Journal, November 8, 1971.

Schapiro, Leonard, et al. "The 24th CPSU Congress." Problems of Communism, July-August 1971.

Schroeder, Gertude E. "Soviet Economic Reform." Problems of Communism, July-August 1971.

Schukin, George S. "The Soviet Position on Trade with the United States." Columbia Journal of World Business 8, no. 4 (winter 1973): 48, 50.

Sukhanov, V., and Shatokhin, E. "Bolshoi Gaz Zapolyar'ya" [Huge Gas of the Arctic]. Izvestia, March 23, 1973, p. 3.

_____. "Rabochiye Trassy Severa" [Workers' Tracks in the North]. Izvestia, March 21, 1973, p. 3.

_____. "Tochka Otchyeta; Zapadno-Sibirskij Kompleks: Opyt i Problemy" [Bench Mark; West-Siberian Compex: Experience and Problems]. Izvestia, March 20, 1973, p. 2.

"Soviets Sold 280 Tons of Gold in '73." Washington Post, March 18, 1974, p. D-7.

"Summary of Long-Term LNG Import Projects." Gas Supply Review, December 15, 1973, p. 22.

Tabor, John K., undersecretary, Department of Commerce. "Achieving a Favorable Balance." International Journal of Research Management 17, no. 4 (July 1974): 9, 10.

Tinbergen, Jan. "Do Communist and Free Economies Show a Converging Pattern?" Soviet Studies, April 1961.

Tracy, Eleanor Johnson. "The Fortune Directory." Fortune, July 1974, pp. 114-15.

"Trade Deficit Deepens." Plain Dealer, August 27, 1974.

Travaglini, Vincent D. "Licensing, Joint Ventures, and Technology Transfer." International Commerce, July 28, 1969, p. 2.

Wett, Ted. "SNG from Coal Involves Big Projects." Oil and Gas Journal, June 25, 1973, pp. 131-34.

Wren, Christopher S. "Russian Oil Profits from the West Climb," New York Times, June 5, 1974, pp. 1, 65.

Zinevitch, A. M., Avenosov. "Soviet Construction Methods for Large
　　Diameter Lines." Pipe Line Industry, February 1973, pp. 42, 44.

UNPUBLISHED MATERIAL

Agishev, A. P., V. G. Vasiliev, and Y. M. Vasiliev. "The Principal
　　Gas-Bearing Areas of the Soviet Union." Paper presented at the
　　12th World Gas Conference, Paris, April 1973, p. 4.

Albright, Raymond J. "Siberian Energy for Japan and the United
　　States." Case study for Senior Seminar in Foreign Policy,
　　Department of State, 1972-73.

Bennsky, George M. "World Trade in LNG: An American Viewpoint."
　　Presentation before the Fourth International Conference on
　　Liquefied Natural Gas at Palace of Nations, Algeria, June 24,
　　1974.

Casey, William J. Export-Import Bank press release, July 9, 1974.

Clark, Ramsey, attorney general. Letter to the secretary of state,
　　May 9, 1967.

Ebel, Robert E. "Gas in the Soviet Union." Paper presented at 48th
　　Annual Fall Meeting of the Society of Petroleum Engineers at
　　Las Vegas, Nevada, October 1-3, 1973.

Export-Import Bank of the United States. Press releases of March 21,
　　1973; September 27, 1973; December 21, 1973; January 18,
　　1974; February 22, 1974; March 22, 1974; and May 21, 1974.

Hardy, Edwin F. "Supplemental Gas Supply Projections." Planning
　　Division, American Gas Association, September 19, 1974.

Kovach, Robert S. "USSR Hard-Currency Balance of Payments."
　　Presentation at the United States Department of Agriculture,
　　November 30, 1973.

Licence, J. V. "Siberia in the Context of World Natural Gas Supplies."
　　Paper prepared for NATO Round Table Meeting, Brussels,
　　January 30, 1974.

Nassikas, John N. "The Role of Liquefied Natural Gas in U.S. Energy Policy." Remarks before the Fourth International Conference on Liquefied Natural Gas at Palace of Nations, Algeria, June 24, 1974.

Schroeder, Gertrude. Remarks made during the East-West Economic Developments Seminar of the Joint Symposium on "Soviet Power and Europe," sponsored by the Association for Advancement of Slavic Studies (AAASS), Washington Chapter, and Institute for Sino-Soviet Studies, The George Washington University, May 10, 1974.

Wheeler, Lyn Floyd. "Export Potential of the U.S. Tool and Die Industry to the Soviet Union through the Sale of Turnkey Plants." D.B.A. dissertation, The George Washington University, 1974.

Whitehouse, Douglas. Presentation made at the Joint Symposium of AAASS at The George Washington University, May 11, 1974.

INTERVIEWS AND CORRESPONDENCE

Albright, Raymond J., vice-president, Planning and Research, Export-Import Bank of the United States, February 1973 to May 1974.

Aksilenko, Valentin P., second secretary, Soviet Embassy, Washington, D.C., November 1972.

Brainard, Dr. Larry, Economic Group, The Chase Manhattan Bank, September 1973.

Cornelius, Frederick, deputy chief, Pipeline Division, Federal Power Commission, May 1974.

Dmitriev, Gennadiy V., chief consultant, Export-Import Division of Machinery and Equipment, Trade Representation of the Union of Soviet Socialist Republics, October 1974.

Farrell, John T., economist, author, contributor, Library of Congress, September 1973 to August 1974.

Jelacic, John E., analyst, United States Tariff Commission, August 1974.

Hardy, Edwin F., manager, Gas Supply, American Gas Association, September 1974.

Kaiser, Mohammed, American Management Association, October 1972.

Koelle, H. Martin, International Monetary Fund, October 1973.

Kovach, Robert S., author, contributor, Library of Congress, November 1973.

Krautler, Charles, public relations officer, Washington Gas Light Company, January 1974.

Lazarus, Steven, deputy assistant secretary for East-West trade, U.S. Department of Commerce, January 1974.

Lee, J. Richard, author, contributor, Library of Congress, August 1973 to August 1974.

Malov, Yuri A., chief of Economic Section, Soviet Embassy, Washington, D.C., October 1974.

Malish, Anton F., Jr., analyst, U.S. Tariff Commission, August 1974.

Mechulaev, Vladimir, second secretary, Soviet Embassy, Washington, D.C., December 1973.

Muse, Ewell H., III, manager of economic planning, Tennessee Gas Transmission Co., April 1974.

Mullin, Robert A., deputy comptroller (International Division), Office of the Comptroller of the Currency, Department of the Treasury, August 1974.

McDonald, Robert P., assistant treasurer, The Chase Manhattan Bank, September 1973 and January 1974.

Porter, Suzanne, East-West Trade Policy Division, U.S. Department of Commerce, August 1974.

Proes, Nevil M. E., president, Texas Eastern LNG, Inc., April 1974.

Ray, Jack H., president, Tennessee Gas Transmission Co., August 1973 to October 1974.

Schroeder, Charles F. A., second vice-president, The Chase
    Manhattan Bank, September 24, 1973.

Reddington, James, energy officer, Department of State, March 1974.

Swett, John C., vice-president, First National City Bank of New York,
    September 1973.

Strishkov, Dr. V. V., mining engineer, Division of Fossil Fuels, U.S.
    Department of the Interior, September 1973 to August 1974.

Tartter, Jean R., deputy director of East-West trade, Department of
    State, August 1974.

Walsh, Daniel S., manager of LNG projects, Texas Eastern Trans-
    mission Corp., September 1973 to August 1974.

Zabijaka, Val, analyst, East-West Trade Division, U.S. Department
    of Commerce, April 1974.

JOSEPH T. KOSNIK is Associate Professor of Finance, Department of Business Administration and Economics, Madison College, Harrisonburg, Virginia. Previously he served for four years as Department Head, Economics, at the Industrial College of the Armed Forces.

Dr. Kosnik has been a consultant to government and private business in the fields of finance and international business. He has had a distinguished career in the U.S. Navy which included six years' service in the Pentagon, where he managed the Navy's Indirect Support Aircraft programs.

Dr. Kosnik received his formal training in engineering, economics, and finance. He earned his B.A. at Notre Dame, a master's degree in Public Administration at Harvard University, and a doctoral degree in Business Administration at George Washington University.

RELATED TITLES
Published by
Praeger Special Studies

EAST-WEST BUSINESS TRANSACTIONS
edited by Robert Starr

INPUT-OUTPUT ANALYSIS AND THE
SOVIET ECONOMY: An Annotated Bibliography
edited by Vladimir G. Treml

THE SOVIET ECONOMY IN REGIONAL PERSPECTIVE
edited by V. N. Bandera and Z. L. Melnyk

THE SOVIET ENERGY BALANCE: Natural Gas, Other
Fossil Fuels, and Alternative Power Sources
Iain F. Elliot

SOVIET INDUSTRIAL IMPORT PRIORITIES: With
Marketing Considerations for Exporting to the USSR
Christopher Stowell, assisted by
Neal Weigel, with chapters by
Edward Maguire and Erast Borissoff

Bibliothèque
Université d'Ottawa

Library
University of Ottawa

JAN 17 '82